U.S. Environmental Protection Agency
Office of Enforcement

National Enforcement Investigations Center
Denver, Colorado

Prepared in cooperation with Environmental
Services Divisions—Regions I through X

Multi-Media Investigation Manual

Government Institutes, Inc.

Government Institutes, Inc., Rockville, Maryland 20850

This manual was prepared in March 1992 by the U.S.
Environmental Protection Agency's National Enforcement
Investigation Center for use within the agency. Government
Institutes determined that it contained information of interest
to those outside EPA; so we are reproducing this material in
order to serve those interested.

This manual is intended as a guide for investigators who
conduct multi-media compliance audits of facilities that
discharge, emit, prepare, manage, store, or dispose of
substances controlled by federal state or local environmental
laws and regulations.

ISBN: 0-86587-300-3

Printed in the United States of America

CONTENTS

INTRODUCTION 1

    PURPOSE AND SCOPE OF THIS MANUAL 5

PROJECT REQUEST/IDENTIFICATION 9

PROJECT TEAM FORMATION 10

    REQUIRED SKILLS AND QUALIFICATIONS 10
    INVESTIGATOR RESPONSIBILITY 12

BACKGROUND INFORMATION REVIEW 13

    FEDERAL/STATE/LOCAL FILE REVIEW 13
    RECONNAISSANCE 14
    EPA DATABASE REVIEWS 14

PROJECT PLAN DEVELOPMENT 16

    PROJECT PLAN 16
    NOTIFICATIONS/SCHEDULING 18

FIELD INSPECTION 21

    DEVELOPING AN INSPECTION STRATEGY 21
    CONDUCTING THE ON-SITE INSPECTION 28

        Entry 28
        Opening Conference 30
        General Facility Tour 34
        Process/Waste Management Evaluation 34
        Document Reviews 37
        Sampling 38
        Closing Conference 39

MEDIA SPECIFIC INVESTIGATION PROCEDURES 41

RESOURCE CONSERVATION AND RECOVERY ACT (RCRA) 41

        Basic Program 41
        Subtitle C - Hazardous Wastes 42
        Subtitle I - Underground Storage Tanks 56
        Subtitle J - Medical Waste Regulations 58
        Pollution Prevention 62

CONTENTS (cont.)

CLEAN WATER ACT (CWA) 64

    Basic Program 64
    Evaluating Compliance 65

CLEAN AIR ACT (CAA) 68

    Basic Program 68
    Evaluating Compliance 70

SAFE DRINKING WATER ACT (SDWA) 73

    Basic Program 73
    Evaluating Compliance 73

TOXIC SUBSTANCES CONTROL ACT (TSCA) 75

    Basic Program 75
    Evaluating Compliance 77

FEDERAL INSECTICIDE, FUNGICIDE, AND
RODENTICIDE ACT (FIFRA) 84

    Basic Program 84
    Evaluating Compliance 84

EMERGENCY PLANNING AND COMMUNITY
RIGHT-TO-KNOW ACT (EPCRA) 87

    Basic Program 87
    Evaluating Compliance 88

COMPREHENSIVE EMERGENCY RESPONSE,
    COMPENSATION, AND LIABILITY ACT (CERCLA) 90
LABORATORY AND DATA QUALITY AUDITS 92

FIELD CITATIONS AND DOCUMENTATION 95

    Field Citations 95
    Other Facility Material 95
    Project Logbook 96
    Chain-of-Custody 96
    Photographs 96

CONTENTS (cont.)

INVESTIGATION REPORT..97

OCCUPATIONAL SAFETY AND HEALTH
      ADMINISTRATION...98

REFERENCES

APPENDICES

A   Summary of Pollution Control Legislation
B   Multi-Media Inspections - Definitions and Training
C   Team Leader Responsibilities and Authorities
D   Types of Information
E   Sources of Information
F   Project Plan and Safety Plan Examples
G   Multi-Media Investigation Equipment Checklist
H   Suggested Records/Document Request
I   Multi-Media Checklists
J   Investigation Authority/Summary of Environmental Acts
K   Evidentiary Procedures for Photographs/Microfilm
L   Key Principles and Techniques for Interviewing
M   Sampling Guidelines
N   Land Disposal Restrictions Program
O   Sources Subpart (40 CFR Part 60) Effective Date of Standard Air
          Pollutants Subject to NSPS
P   TSCA Forms
Q   TSCA Sections 5 and 8 Glossary of Terms and Acronyms

FIGURE

1   Project Phases..7
2   Multi-Media Investigative Approach...23
3   Multi-Media Inspection Feedback Loops....................................26

TABLE

Permitted vs. Interim Status...................................47

Mention of commercial names is for example purposes and does not constitute endorsement by EPA. Additionally, this manual supplements, not supersedes, Standard Operating Procedures, Policy and Procedures Manuals, and any other EPA guidance.

INTRODUCTION

INTRODUCTION

This manual is intended as a guide for investigators who conduct multi-media environmental compliance investigations of facilities that discharge, emit, prepare, manage, store, or dispose of pollutants controlled by Federal, State, or local environmental laws and regulations. Investigative methods are presented that integrate the enforcement programs for air, water, solid waste, pesticides, and toxic substances. This manual describes general activities and functions of multi-media investigations, and provides information on special features of specific media and associated statutes. This manual is intended to supplement the various media-specific investigation guides listed in the reference section.

Multi-media compliance investigations are intended to determine a facility's status of compliance with applicable laws, regulations, and permits.

The environmental laws which EPA administers and enforces are summarized in Appendix A. Emphasis is given to identifying violations of regulations, permits, approvals, orders and consent decrees, and the underlying causes of such violations. Investigators should thoroughly identify and document violations and problems that have an existing or potential effect on human health and the environment.

All inspections can be grouped into four categories of increasing complexity, moving from Category A (a program-specific compliance inspection) to Category D, (a complex multi-media investigation) depending upon the complexity of the facility and the objectives of the investigation. Factors in categorizing the investigation include the complexity of pollution sources, facility size, process operations, pollution controls, and the personnel and time resources which are required to conduct the compliance investigation. The four categories of investigations are described below [Appendix B]:

<u>Category A</u>: Program-specific compliance inspections, conducted by one or more inspectors. The objective is to determine facility compliance status for program-specific regulations.

Category B: Program-specific compliance inspections (e.g., compliance with hazardous waste regulations), which are conducted by one or more inspectors; however, the inspector(s) screen for and report on obvious, key indicators of possible noncompliance in other environmental program areas.

Category B multi-media inspections have limited, focused objectives and are most appropriate for smaller, less complex facilities that are subject to only a few environmental laws. The objective is to determine compliance for program specific regulations and to refer information to other programs based on screening inspections.

Category C: Several concurrent and coordinated program-specific compliance investigations conducted by a team of investigators representing two or more program offices. The team, which is headed by a team leader, conducts a detailed compliance evaluation for each of the target programs.

Category C multi-media investigations have more compliance issues to address than the Category B inspection and are more appropriate for intermediate to large facilities that are subject to a variety of environmental laws. The objective is to determine compliance for several targeted program-specific areas. Reports on obvious, key indicators of possible noncompliance in other environmental program areas are also made.

Category D: These comprehensive facility evaluations address not only compliance in targeted program specific regulations, but also try to identify environmental problems that might otherwise be overlooked. The initial focus is normally on facility processes to identify activities (e.g., new chemical manufacturing) and byproducts/waste streams potentially subject to regulation. The byproducts/waste streams are traced to final disposition (on-site or off-site treatment, storage, and/or disposal). When regulated activities or waste streams are identified, a compliance evaluation is made with respect to applicable requirements.

The investigation team, headed by a team leader, is comprised of staff thoroughly trained in different program areas. For example, a large industrial facility with multiple process operations may be regulated under numerous environmental statutes, such as the Clean Water Act (CWA), Clean Air Act (CAA), Resource Conservation and Recovery Act (RCRA), Toxic Substances Control Act (TSCA), Comprehensive Environmental Response, Compensation and Liability Act (CERCLA), and the Federal Insecticide and Rodenticide Act (FIFRA). The on-site investigation is conducted during one or more time periods, during which intense concurrent program-specific compliance evaluations are conducted, often by the same cross-trained personnel.

Category D multi-media investigations are thorough and, consequently, resource intensive. They are appropriate for intermediate to large, complex facilities that are subject to a variety of environmental laws. Compliance determinations are made for several targeted program-specific areas. Reports on possible noncompliance are made for other program areas.

Generally, all investigations will use essentially the same protocols, including pre-inspection planning, use of a project plan, sampling, inspection procedures, and final report. The major difference will be in the number of different regulations Categories C and D investigations address.

The multi-media approach to investigations (which is described in Categories C and D) has several advantages over a program-specific inspection, including:

- A more comprehensive and reliable assessment of a facility's compliance with fewer missed violations

- Improved enforcement support and better potential for enforcement

- A higher probability to uncover/prevent problems before they occur or before they manifest an environmental or public health risk

- Ability to respond more effectively to non-program specific complaints, issues, or needs and develop a better understanding of cross-media problems and issues, such as waste minimization

The success of a multi-media investigation program is contingent upon a good managerial system and the support of upper management. Since these investigations will often be conducted at larger facilities, adequate resources (time and personnel) must be provided. Good communications during the planning phase are essential to define the scope of work and each team member's role in the inspection. Communications could also include state officials since state inspectors might also participate as team members. Often, because of the extent of the state's knowledge of the facility and its problems, state involvement is critical to the success of the investigation. Similarly, coordination with other Federal or local agencies needs to be addressed, as necessary.

Branch Chiefs and Section Chiefs are important in implementing the multi-media inspection program and identifying areas of responsibility and accountability. Some of their duties include:

- Identify team leaders
- Form investigation teams
- Provide access to training and other means necessary to develop multi-media investigation expertise
- Participate in targeting investigations

- Ensure that team activities both internal and external to their Divisions are coordinated
- Market multi-media investigations to programs
- Oversee the preparation of a site-specific project plan and safety plan prior to the investigation
- Provide managerial support while teams are in the field
- Ensure quality of final reports

The roles and areas of responsibility and accountability of other managers, technical staff, and team leaders must be defined. Participants need to identify and agree on what evidence is needed and the scope of work to be conducted. Next, a project plan and safety plan that outlines the desired objectives and safety considerations must next be prepared by the team leader. Other responsibilities for the technical staff, which often mirror or complement those of managers, are as follows:

- Contact state counterparts
- Assist in investigation preparation, including logistical considerations
- Coordinate activities internal and external to their Division
- Provide legal support for obtaining warrants when necessary
- Provide training for investigators
- Prepare reports
- Distribute reports and followup for multi-media enforcement

PURPOSE AND SCOPE OF THIS MANUAL

Multi-media investigations are carried out in response to specific requests from the EPA program offices, legal staff, or state environmental offices. All investigations will result in a written report that documents non-compliance or other areas of concern identified during the investigation. Report guidelines for documenting a multi-media investigation are discussed later in this document.

This manual provides guidelines for conducting Categories C and D multi-media investigations, as well as, suggests principles and procedures which will also apply to Categories A and B or single media investigations. Moreover, this manual identifies multi-media objectives and also focuses on specific environmental laws and associated statutes.

The manual's organization follows the steps involved in a multi-media investigation beginning with the project request and leading ultimately to enforcement case support.

Multi-media investigations are conducted as a series of tasks or phases, which usually include:

- Project request/identification of objectives
- Project team formation
- Background information review
- Project Plan preparation
- On-site field inspection
- Report preparation
- Enforcement case support

These phases are depicted in Figure 1 and discussed in detail in the following sections. Each phase is not necessarily a discrete step to be completed, in order, before initiating the next phase. The background information review, for example, usually continues through enforcement case support.

Where established policies and procedures do not exist, common sense, professional judgment, and experience should be applied. Investigators need to collect valid, factual information and supporting data to document violations adequately. The documentation must be admissible as evidence in any subsequent enforcement action.

Figure 1
PROJECT PHASES

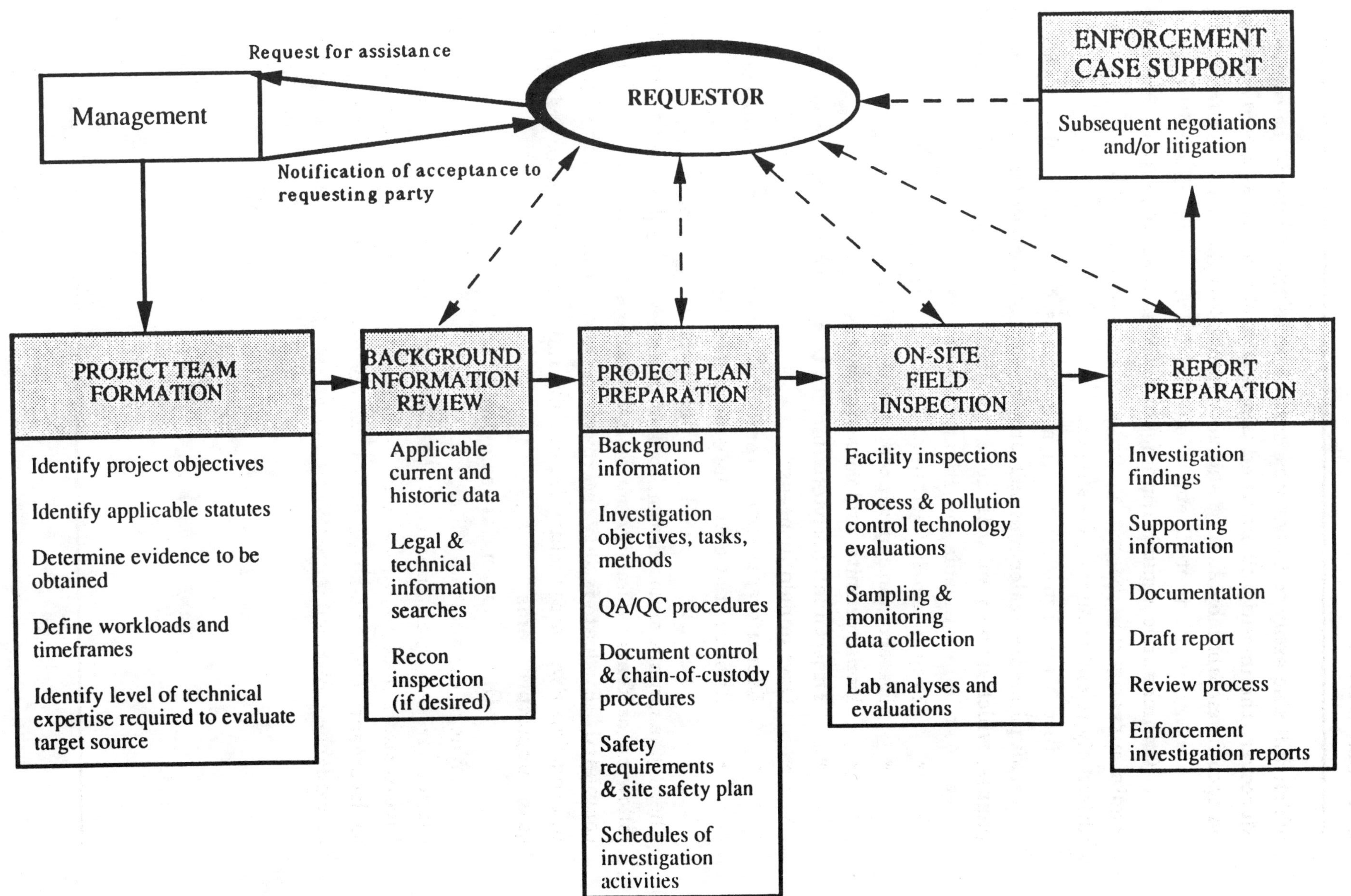

Each investigation should be conducted as though it will be contested in court. The investigation and all supporting evidence and documentation may be contested by highly skilled defense counsel as unprofessional, inaccurate, misinterpreted, etc. Because the Agency's case will depend heavily on the investigative findings, the investigation must be complete and accurate.

Since a multi-media investigation by its very nature probes into a facility's processes under multiple environmental regulations, it provides a highly effective way of looking at a non-complying facility. Overall objectives of a multi-media investigation include:

- Determine compliance status with applicable laws, regulations, permits, and consent decrees
- Determine ability of a facility to achieve compliance across all environmental areas
- Identify need for remedial measures and enforcement action(s) to correct the causes of violations
- Evaluate a facility's waste producing, treatment, management, and pollution control practices and equipment
- Evaluate facility self-monitoring capability
- Evaluate facility recordkeeping practices
- Evaluate facility waste minimization/pollution prevention programs
- Obtain appropriate samples

Project Request/ Identification

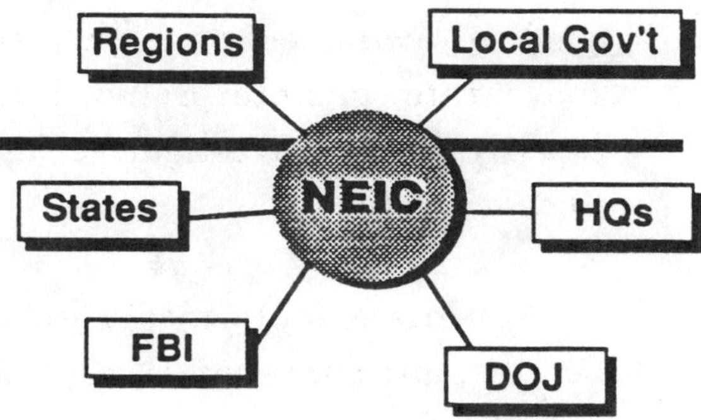

The success of a multi-media investigation depends on thorough, up-front planning. Coordination with all interested and knowledgeable parties [i.e., Region, State, National Enforcement Investigations Center (NEIC), investigation team members, supervisory staff] is essential to ensure an efficient and thorough investigation. Each program office should be contacted to assure that all interested parties are aware of the planned multi-media investigation. Furthermore, depending upon State/EPA agreements, the State should (may) be notified of the pending investigation. All concerned parties should be involved as soon as possible to facilitate coordination. Coordination with other Federal agencies, such as Occupational Safety and Health Administration (OSHA) or Corps of Engineers, may be needed in joint Federal investigations.

There must be agreement on the part of program and support offices and Regional Counsel, and perhaps the State as well, on the overall scope of work for a particular investigation. The scope must be clearly communicated to managers and the investigators and stated in the project plan. Clearly defined objectives are critical to the success of any investigation; the objectives should be well defined at the time of the initial request. At the time the project plan is developed, the objectives will be further refined (see section titled Project Plan Formation). Communication among the involved parties should be the initial step of pre-investigation planning but should also continue throughout the investigation, because of unanticipated events often associated with field work.

PROJECT TEAM FORMATION

REQUIRED SKILLS AND QUALIFICATIONS

Each multi-media investigation team should be composed of qualified field inspectors. Each team member should bring special program expertise and experience and must be well trained in most facets of conducting a field investigation, including sampling.

Most of the investigators on the team, including the team leader, should be current field investigators who already possess most of the necessary skills and qualifications. However, flexibility in team selection must exist in order to use skills in the organization and ensure that the expertise is represented for a given situation. There may be circumstances where a permit writer, hydrogeologist, toxicologist, or some other special discipline will be needed on the team.

The team leader has overall responsibility for the successful completion of the multi-media investigation. (Team leader responsibilities and authorities are presented in Appendix C.) In addition, other investigators may be designated as leads for each of the specific media/programs that will be addressed. These individuals may work alone or have one or more inspectors/samplers as assistants, depending on workloads and training objectives. However, all investigation team members should report directly to, and be accountable to, the team leader.

Some of the more important skills and qualifications that are necessary for team members, are as follows:

- Knowledge of the Agency's policies and procedures regarding inspection authority, entry procedures/problems, enforcement actions, legal issues, and safety

- Thorough familiarity with sampling equipment, quality assurance (QA) requirements for sample collection, identification and preservation, and chain-of-custody procedures

- Knowledge of manufacturing/waste producing processes, pollution control technology, principles of waste management, flow measurement theory and procedures, and waste monitoring techniques/equipment

- Investigatory skills including ability to gather evidence through good interviewing techniques and astute observations

- Up-to-date experience in conducting compliance inspections

- Communication skills

- Basic understanding of the procedures for obtaining administrative warrants, including preparation of affidavits, technical content of the warrant application, and warrant and procedures for serving a warrant

- For each of the areas addressed in the multi-media investigations, there should be at least one team member trained in that area. Furthermore, at least one team member should have considerable knowledge of laboratory (analytical) methods and quality assurance (QA) requirements, if a laboratory evaluation is to be conducted.

EPA Order 3500.1 sets forth specific training requirements for any EPA investigator who is leading a single media investigation. These training requirements include both general inspection procedures and media specific procedures. While an individual leading a multi-media investigation may not have had the media-specific training for each media covered during that multi-media investigation, the team leader should have the media specific training for at least two of the media. In addition, the team leader should have experience and skills in the following areas.

- Leadership

- Project management

- Lead for inspections in more than one program/media

- Multi-media investigations

INVESTIGATOR RESPONSIBILITY

Investigators must conduct themselves in a professional manner and maintain credibility. A cooperative spirit should be cultivated with facility representatives, when possible. All investigators should maintain a sensitivity to multi-media issues and implications and freely discuss, with other members of the team, observations/findings relating to one or more programs (or cross-program lines). Moreover, investigators must remember to adhere to the project plan,[*] as well as both the Agency's and the facility's safety plans or requirements.

Investigators should restrict their on-site activities to the normal working hours of the facility, as much as possible. Investigators will need to keep abreast of specific program regulations and should also coordinate, as necessary, with other EPA and State inspectors and laboratory staff (if samples will be collected). The investigation team should implement appropriate field note taking methods and proper document control procedures, particularly when the company asserts a "confidential" claim. Investigators must assume that important documents (e.g., project plan, safety plan, and logbooks) are not left unattended at the facility. Sensitive discussions do not take place in front of facility personnel or on company telephones.

---

[*] *Field conditions may dictate plan modifications. Reasons for modifications should be thoroughly documented.*

BACKGROUND INFORMATION REVIEW

FEDERAL/STATE/LOCAL FILE REVIEW

The investigation team must collect and analyze available background information in order to better plan and perform the multi-media investigation. The objective of the review is to allow EPA staff to (1) become familiar with the facility, (2) clarify technical and legal issues prior to the inspection, (3) use resources wisely, and (4) provide information to allow each inspector to develop a list of questions to be answered and documents to be obtained during the on-site inspection. For example, the investigators should understand the facility's process(es) to the extent possible and know where past problems have occurred, based on file/data reviews. Much of the total time spent on an investigation should be spent on planning and preparing for the investigation. This will prevent classic oversights such as being on the road and not knowing where the facility is, or walking past the operation that received a Notice of Violation in five previous inspections. Investigators should check with the program staff (Federal/State/local, etc.) to gain as much knowledge as possible about the site. Federal/State/local file reviewers should pay particular attention to the following:

- Permits and permit applications
- Process and wastewater flow charts
- Prior inspection reports
- Enforcement documents including Administrative Orders, Complaints, Consent Decrees/Agreements, Notices of Noncompliance (NONs), Deficiency Notices, Compliance Schedules, Cease and Desist Orders, Closeout Documents, Notices of Violations, etc.
- Facility responses to all of the above
- Facility records, reports, and self-monitoring data
- QA documentation
- Exemptions and waivers
- Maps showing facility layout and waste management/ discharge sites
- Records of citizen complaints

- Consultant's reports
- Potential cross-program issues
- Annual reports
- Hazardous waste manifests
- Spill reports

A more detailed list of types of information which may be acquired and reviewed is presented in Appendix D.

RECONNAISSANCE

A reconnaissance inspection of the facility may be conducted in conjunction with gathering background information from State and local files. Administrative details and logistics are usually discussed during a reconnaissance that will help the on-site inspection proceed more efficiently. A reconnaissance inspection is particularly important if a complex facility is being investigated or if the facility has never been inspected by the team leaders. At least the team leader should participate in the reconnaissance. No reconnaissance is conducted if the investigation will be unannounced, or if the team has extensive knowledge of the facility (see Project Plan Formation Section).

EPA DATABASE REVIEWS

Additional facility background material should be obtained from EPA databases. (Acronyms are defined in Appendix E.) At a minimum, the inspectors should use the following:

- TRIS (provides facility data on past releases of toxic/hazardous substances to the environment, as required by Section 311 of SARA)

- DUNS Market Identifiers: Commercial systems that tracks the owners and financial information for publicly- and privately-owned companies in the U.S.

- PCS (provides CWA/NPDES permit related information, DMR data, receiving stream data, some enforcement related material, and inspection history for "major" wastewater discharges)

- RCRIS/HWDMS (provides RCRA-related information on a facility such as location, hazardous waste handled, inspection history, nature of past violations, and results of enforcement actions)

- FTTS (provides TSCA-related information on a facility such as inspection history, and case development information, including violations, and types/results of enforcement actions)

- FINDS (EPA database that identifies regulations applicable to the target facility, including some related to compliance/ enforcement issues)

- CERCLIS (Superfund's national database system provides information on CERCLA sites)

- AFS/AIRS (the Air Compliance Program's national database system provides air compliance information for major sources)

A more extensive list of sources of information, including both computer databases and other sources, is presented in Appendix E. Following file/data reviews, the investigation team may prepare a fact sheet for the facility along with a list of questions that need to be answered either before or during the on-site actual investigation.

PROJECT PLAN DEVELOPMENT

PROJECT PLAN

A site-specific project plan should be developed for all multi-media inspections. Each project plan should reflect the requirements/scope of work associated with each individual facility. The plan describes the project objectives and tasks required to fulfill these objectives. In addition to methods, procedures, resources required, and schedules, a safety plan is included as an appendix and identifies potential site safety issues, procedures, and safety equipment [Appendix E].

Generally, a draft project plan is prepared to give all involved parties/regional management an opportunity to review the planned project activities and schedule. The team leader, with the assistance of other investigators, is responsible for preparing the site-specific project plan. After agreement on the draft is reached, the plan should be finalized as soon as possible. It must be available before the on-site inspection starts. A comprehensive project plan provides a means for informing all involved parties of the proposed activities and helps ensure an effective multi-media investigation; team members must be familiar with the project plan.

The following generally form the outline for the project plan:

<u>Objectives</u> - This is probably the most important part of the project plan and should be well defined, complete, and clear. The objectives should also have been discussed and agreed upon by all appropriate management personnel. The objectives define what the investigation is to accomplish (e.g., to assess environmental compliance with the regulations that apply to the source--water, air, *et al.*).

<u>Background</u> - Discusses, in general, facility processes and, based on available information, identifies laws, regulations, permits, and consent decrees applicable to the facility.

<u>Tasks</u> - The plan defines tasks for accomplishing the objectives and spells out procedures for obtaining the necessary information and

evaluating facility compliance. The tasks usually involve an evaluation of process operations, pollution control/treatment and disposal practices, operation and maintenance practices, self-monitoring, recordkeeping and reporting practices, and pollution abatement/control needs. Tasks will probably be sequenced, based on: facility factors, investigation objectives, logistical factors, constraints imposed by the company, and complexity and overlap of regulatory programs (see section on Field Inspections).

<u>Methods/Procedures</u> - The plan provides or references policies and procedures for document control, chain-of-custody, quality assurance, and handling and processing of confidential information. Specific instructions for the particular investigation may be provided.

<u>Safety</u> - A safety plan attached to the project plan identifies safety equipment and procedures which the investigation team must follow [Appendix F]. The safety procedures and equipment are typically the more stringent of EPA or company procedures. EPA procedures are documented in EPA Transmittal 1440 - Occupational Health and Safety Manual dated March 18, 1986. Additional safety issues for extensive or prolonged investigations, if necessary, should also be addressed in the plan.

<u>Resources</u> - The plan describes special personnel needs and equipment requirements. As noted earlier, experienced and knowledgeable personnel shall compose the investigation team. An example of an equipment list is presented in Appendix G.

<u>Schedules</u> - The plan usually provides general schedules for investigation activities. This information is important to the team members as well as Headquarters, Regional, and/or State officials who requested the project. The dates for (a) starting and finishing the field activities, (b) analytical work, and (c) draft and final reports should be established and agreed upon by the participants.

The project plan will serve as the basis for explaining inspection activities and scheduling to facility personnel during the opening conference. The company may be provided general details but should not be provided with a copy of the project plan; it is an internal document and usually considered an enforcement confidential document. (The company may get a copy of the plan by court order.) Because conditions in the field may not be as anticipated, the project plan is always subject to modification and so marked. Any deviations from the plan should be well thought out, approved by the team leader, and if appropriate, discussed with senior management or the laboratory (if monitoring/sampling requirements change) and well documented.

NOTIFICATIONS/SCHEDULING

Notification of the inspection, normally by telephone, should be given to the facility unless the inspection is to be unannounced. There are advantages to both announced and unannounced inspections. Some of the advantages to announced inspections include assuring that the right people are available, the processes of interest are operating, and the necessary documents are available. A major consideration of announced inspections is that the company may be able to alter operations to conceal violations. Following the telephone notification, it may be necessary to prepare a more formal notification letter or notification form that is served when entering the facility.

If a letter is prepared, it should cite the appropriate inspection authorities, the general areas to be covered, and special informational needs/requests. By citing broad EPA authorities, the investigation will not be restricted if investigators need to pursue additional areas based on field observations. The notification should specify records to be reviewed/ collected/copied, address safety/security issues, and include any questions that need answering to help facilitate the investigation. These issues and questions can also be addressed during a reconnaissance inspection, if desired, or through telephone conversations with appropriate facility personnel.

Typical information requested in a notification letter may include the following:

- Raw materials, imports, intermediates, products, byproducts, production levels

- Facility maps identifying process areas, discharge and emission points, waste management and disposal sites

- Flow diagrams or descriptions of processes and waste control, treatment and disposal systems, showing where wastewater, air emission, and solid waste sources are located

- Description and design of pollution control and treatment systems and normal operating parameters

- Operations and maintenance procedures and problems

- Self-monitoring reports and inventories of discharges and emissions

- Self-monitoring equipment in use, normal operating levels, and available data

- Required plans, records, and reports

Appendix H identifies specific documents/records by statute which might be requested. Each regional office should decide if and when state regulatory officials will be notified of the investigation and who will make the contact. By reviewing State files, EPA will have, in effect, notified the State of its intention to inspect this facility. The State should be requested to allow only EPA to notify the facility regarding the multi-media inspection. If sampling is anticipated, the laboratory should be notified as soon as possible, and informed as to when samples will arrive and the approximate analytical work load. The project plan, which is reviewed by laboratory personnel, should also identify analytical support required.

The investigation should be scheduled at a time mutually agreed upon by all participants. Sufficient time should be allotted to conduct a thorough investigation. Appropriate travel arrangements should be made as soon as possible.

FIELD INSPECTION

Once the project plan is completed, the team's focus shifts to the on-site portion of the investigation. This section first addresses developing a site-specific inspection strategy for evaluating processes and regulated waste management activities, then discusses on-site activities from entry through the closing conference.

The primary objectives of the field inspection are to determine whether the facility is complying with environmental regulations, permits, etc., and to determine if facility activities are creating environmental problems. The investigation team should also determine if the facility has environmental management controls in place to maintain regulatory compliance (i.e., a system for becoming aware of regulatory requirements, then implementing appropriate compliance actions) and whether the controls are working. By satisfying these objectives, areas of non-compliance, environmental problems, and insight into root causes can be identified during the investigation. The information will be useful later in followup actions.

DEVELOPING AN INSPECTION STRATEGY

Inspection planning includes formulating a strategy to ensure that information is obtained from the company in a logical, understandable manner. This applies to both the process and compliance evaluations, and the environmental management control evaluation. To formulate an effective strategy, knowledge of general facility operations, waste management procedures, and applicable regulations is critical. Much of this information should have been obtained during the background information review and inspection reconnaissance. This section first discusses strategy development, then presents an example.

The process evaluation strategy to sequence inspections for major facility operations and waste management activities may be based on:

- Facility factors such as size, complexity, waste producing potential, and waste characteristics

- Administrative factors such as the priority of inspection objectives (i.e., which compliance evaluations are the most important)

- Logistical factors such as personnel availability, operating schedules, and assignment schedules

- Constraints imposed by the company such as limitations on the number of inspection teams that can operate on-site concurrently

The final strategy usually involves prioritizing the processes and waste management activities, in consideration of all these factors, then systematically moving from the beginning to the end of a process with emphasis on regulated waste stream generation and final disposition. The strategy should be somewhat flexible so that "mid-course corrections" can be made.

The compliance evaluations also need to be "sequenced" in a similar manner to progress, generally, from the most to least time-consuming regulatory program. Personnel training and availability, and other logistical factors may result in a combining of compliance evaluations. Figure 2 (Investigative Approach) illustrates a sequence of compliance evaluations where the initial focus was on RCRA, then CWA, etc. RCRA is often chosen as the initial law to focus on because of the close relationship between process evaluations and generator requirements. A quick visual inspection of hazardous waste storage areas and PCB transformers is often conducted early in the inspection. Compliance with regulatory programs that principally involve records reviews, such as TSCA (Sections 5 and 8) and EPCRA are usually scheduled later in the inspection or elsewhere, as time permits.

Figure 2
Multi-Media Investigative Approach

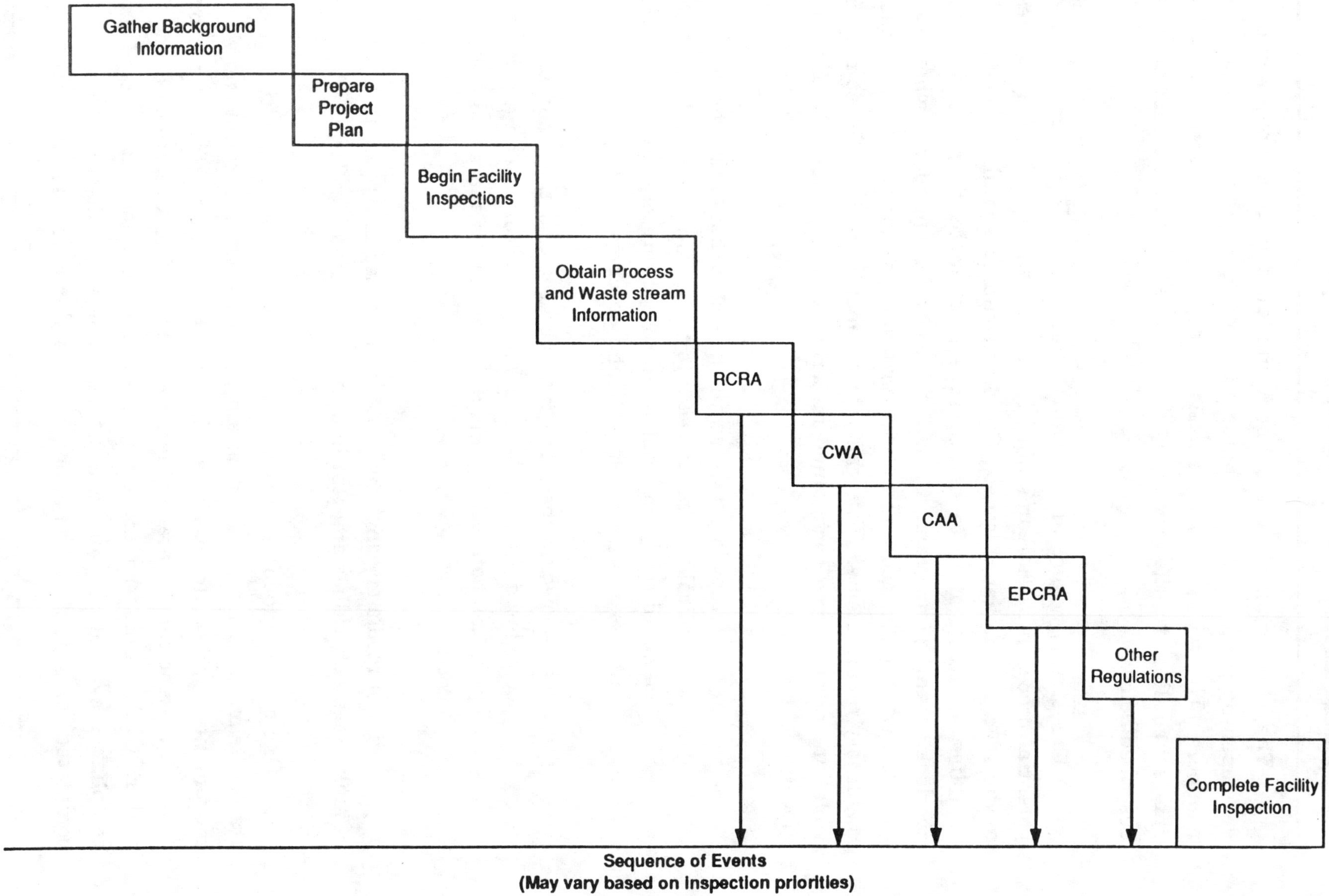

The strategy for process and compliance evaluations should be developed by the project coordinator and discussed with inspection team members. This will serve as the basis for explaining inspection activities and scheduling to the company during the opening conference, as described below.

The strategy may also include checklists. Some may address potential process wastestreams to be looked for, while others may address media-specific compliance issues. Checklists can be a vital component of a compliance investigation to help ensure that an investigator does not overlook anything important. Checklists serve as a reminder of what needs to be asked or examined and to help an investigator remember the basic regulatory requirements. They can provide another means of documenting violations or supplying background material to judge potential violations, however:

- An inspector must never fill out a checklist blindly or too mechanically. The answers to the questions should not be based solely on what the facility representatives say, but also on what the investigator observes.

- Media-specific checklists may be used and they may be completed by the lead investigator for each given program, both during and after the facility tours and the document review phases.

A list of media-specific checklists is presented in Appendix I. Copies of multi-media checklists are kept in a three-ring binder at NEIC.

One of the unique benefits of the Category D approach with a cross-trained team is that information obtained on processes, material and waste movements, and scale of operations can later aid in focusing other program-specific compliance evaluations, such as TSCA (Sections 5 and 8) and EPCRA. Like the project phases, the sequence of process and compliance evaluations should not be considered as discrete steps to be completed, in order. Information obtained during subsequent program-specific evaluations may also provide new information regarding compliance in a program area already addressed or indicate a need to inspect a process/support operation not previously identified [Figure 3] This iterative process is pursued until the inspection objectives have been accomplished.

At larger facilities, multiple site visits coordinated by the team leader may be necessary and desirable for completing the inspection. This approach can lead to a better inspection because of the opportunity to review information obtained in the office, then refine the inspection/strategy to "fill in the gaps" during a subsequent site visit.

An inspection strategy example for a typical facility is presented below:

INSPECTION STRATEGY EXAMPLE

1. SITE OVERVIEW/PROCESS OVERVIEW
2. WINDSHIELD TOUR
3. SPLIT INTO TEAMS FOR DETAILED PROCESS REVIEW AND PROCESS AND LABORATORY INSPECTIONS
4. RECORD REVIEW PERIOD
5. FOLLOW-UP INTERVIEWS

After the opening conference to explain inspection activities, the company would be requested to provide an overview of facility operations to the entire team. A general "windshield" site tour usually follows the overview presentation. Next, process operations would usually be described in some detail; the order typically parallels the flow of raw materials and intermediate products toward subsequent processes and the final product(s). During these discussions, waste streams and respective

Figure 3
Multi-Media Inspection Feedback Loops

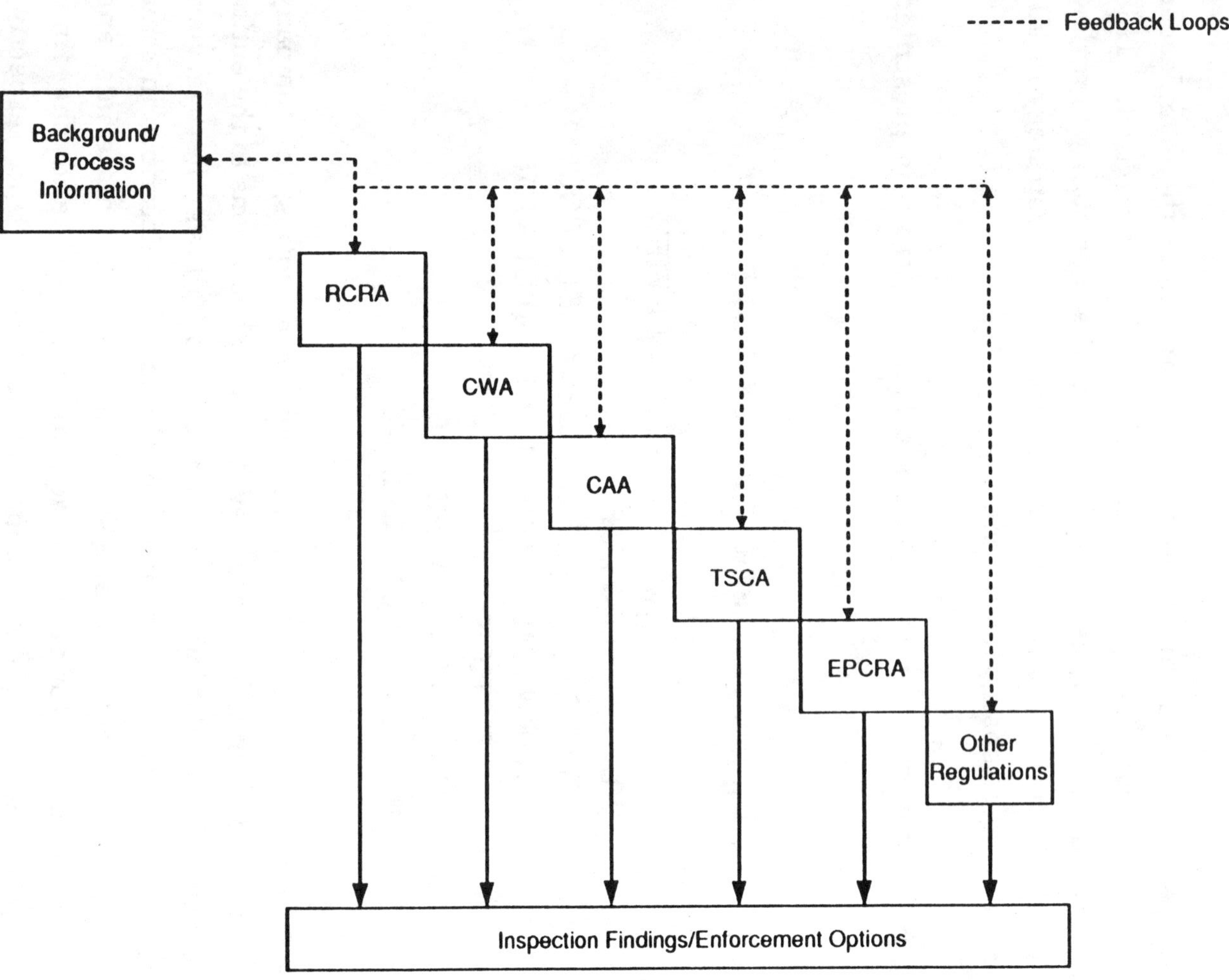

management procedures and related control equipment should be identified.

Process inspections to verify the information presented and discover "missing" details are then conducted. These may be done after each process or group of processes is described. The RCRA inspection begins while touring the processes by identifying any waste generation and accumulation areas. The presence of wastewater sumps, separators, or trap tanks in or near the process building may also result in initiating the CWA inspection.* The example plan indicates that the inspection team subdivided before proceeding with the detailed process descriptions. This is because the people gathering process information were beginning some of the compliance evaluations; other team members could begin concurrent evaluations such as the laboratory inspection.

Inspection of waste management units may be interspersed with process inspections depending on their location and facility complexity; a primary consideration is the logical flow of information. Logic or the physical situation may dictate that a waste stream be followed to final disposition for a particular process. Time must also be scheduled to review and copy relevant records, then for additional interviews to answer questions about the records.

The evaluation of environmental management controls is blended into the process and compliance evaluations discussed above. Investigation team members should allow facility personnel to explain their operations until the management system is understood. Document the management system with narrative notes, gather copies of all documents used in the processes, and formulate flowcharts to illustrate decision responsibilities, accountabilities, and process matrices. Environmental regulations, permits, etc., are the standards, and the internal management systems

---

* *The CWA regulates outfalls from wastewater treatment plants and other point discharges. If the sumps, separators, and trap tanks are used to manage hazardous waste, they may be exempt from RCRA if they discharge to an on-site treatment plant that has a CWA permit. Part of the CWA inspection, therefore, is identifying all treatment units related to the permitted (or unpermitted) outfall.*

and procedures are the controls established to direct compliance or conformance to the standards.

The system should be tested by tracking information from the internal management systems through the regulated activity locations (i.e., follow known events through entire processes). For example, if a facility's environmental coordinator states that a particular waste analysis plan (a RCRA requirement) is being used at the facility, the investigation team would "test" the system by verifying that personnel at the waste receiving station and laboratory (1) had a copy of the plan, (2) were familiar with it, and (3) were following it.

Finally, continuing communication between team members is a key to successful strategy implementation. The team leader should encourage daily team meetings (usually in the evenings) to discuss findings and observations made during the day. Ensuing discussions may help to clarify any troublesome issues or open up new avenues for investigation.

CONDUCTING THE ON-SITE INSPECTION

The field portion of a multi-media investigation involves entering the facility and conducting an on-site inspection. The following discusses several key inspection activities including: entering the facility, conducting an opening conference, systematically gathering pertinent information while on-site, and discussing findings in a closing conference.

Entry

Entry into a facility to conduct a multi-media inspection is usually a straightforward process where the team leader notifies a guard and/or receptionist that he/she wants to meet with the designated facility environmental contact. The environmental contact is notified and the team is then escorted into the facility to begin the inspection, usually with an opening conference, where credentials are officially presented.*

Many facilities request inspection team members sign a visitor's log or some other document that will provide a written list of inspectors to the company. This is acceptable so long as there is no waiver of liability or restriction of inspection statements contained on the document being signed (check both sides of the paper for such statements, or the first page of a bound sign-in log). Liability waivers must not be signed; signature documents containing any statements that appear to limit inspection activities should be first discussed with Agency attorneys (if in doubt, consult), or sign in only on blank sheets.

Occasionally, entry is denied, usually in situations where the inspection is unannounced or enforcement action is pending (e.g., outstanding Notice of Violation, ongoing Administrative Order negotiations, etc.). Consequently, the team leader needs to know how to deal with denial of entry situations.* Whenever entry consent is denied (or withdrawn during the course of the inspection), the team leader should explain the Agency authority to conduct the investigation and verify that the authority is understood by the facility representative. If the person persists in denying entry or withdrawing consent, the team leader needs to fully document the circumstances and actions taken; this includes recording the name, title, and telephone number of the person denying entry or withdrawing consent. Inspection team members must never make threatening remarks to facility personnel. The team leader should then contact his/her supervisor and Agency legal counsel.

If the team leader suspects that a warrant will be necessary before entry is attempted, then actions to obtain a warrant (e.g., get attorney assigned to the project and prepare draft affidavits) should be initiated well

---

*	*In an unannounced inspection, it may be beneficial to immediately go to the regulated areas of concern (drum storage, etc.) to conduct an inspection before the facility has time to make changes.*

*	*An attempt by the company to unreasonably limit legitimate team activities during the inspection is tantamount to denial of entry and should be treated as such.*

in advance to minimize the time between arrival at the facility and entry.[*] This would lessen the opportunity for the facility to take drastic, last-minute corrective actions (e.g., improve "housekeeping") and thereby give the investigators a false impression. If a warrant becomes necessary, it is obtained from a local magistrate or judge; an EPA special agent, a Federal Marshal, or sheriff will be needed to serve the warrant. The team leader must be familiar with the warrant provisions and be aware of both opportunities and constraints imposed on the investigation.

Opening Conference

The opening conference is held to advise facility personnel of the investigation objectives, and to discuss logistics and scheduling of inspection activities. An important aspect of this meeting is to set the "proper tone" with facility personnel (i.e., encourage cooperation). A typical conference agenda includes:

- Introduction of investigators and presentation of credentials (be prepared to cite investigation authorities - Appendix J)

- Description of investigation objectives:

  The investigation objectives have been generally identified in the project plan; however, the project plan should not be shown to the company. As noted earlier, this is an internal document and not to be released by the field investigation team to the company. Additionally, during the discussion of investigation objectives, the investigators should take care not to limit the investigation if as a result of their findings, a new objective becomes apparent.

---

[*] *A warrant is only one of several legal vehicles that should be considered. A TSCA subpoena, issued to a high-ranking corporate official, was used successfully in one instance as a vehicle to gain consensual entry; the limitations imposed by a warrant were avoided.*

The purpose of identifying the investigation objectives to the company is to enable the company to identify what people and what documents are necessary to assist in the investigation.

- Description of investigation procedures and personnel needed (develop schedule of events)

Let the management know approximately how long the investigation will take so they can assist with the least interruption of their regular schedules. This is often as long as it takes until we get the information requested from the facility.

While the government has a right to inspect at any time during normal working hours, it is appropriate to give some consideration to the needs of the facility. For example, perhaps giving company personnel one half hour in the morning to get their business in order would be beneficial and would win "good will" for the investigators.

- Presentation of inspection notices/forms

- Discussion of prospective sampling and whether company splits will be made available

- Discussion of safety issues including the company's safety requirements [Appendix F]

The government investigators may not have the same restraints as facility personnel. However, it is prudent to determine what safety requirements the facility personnel have to follow and to follow those if they are more stringent than the government requirements. Of particular importance is to determine emergency signals and escape routes if a plant emergency occurs. Commonly, investigators may attend a short safety briefing and be asked to sign that they attended that briefing. It is all right to sign an acknowledgement that a safety briefing was attended; it is not all right to agree to anything else or to relinquish any rights CHECK WITH THE GOVERNMENT ATTORNEYS IF THERE ARE ANY QUESTIONS.

- Discussion on how photographs will be taken

Photographs are used to prepare a thorough and accurate investigation report, as evidence in enforcement proceedings and to explain conditions found at the plant. The facility, however, may object to the use of cameras in their facility and on their property. If a mutually acceptable solution cannot be reached and photographs are considered essential to the investigation, Agency supervisory and legal staff should be contacted for advice.

Facility personnel may also request that photographs taken during the visitation be considered confidential, and the Agency is obliged to comply, pending further legal determination. Self-developing film, although often of less satisfactory quality, is useful in these situations. A facility may refuse permission to take photographs unless they can see the finished print. Duplicate photographs (one for the investigator and the other for the facility) should satisfy this need. When taking photographs considered TSCA Confidential Business Information (CBI), self-developing film eliminates processing problems; otherwise, the film processor must also have TSCA CBI clearance. Note, however, that some self-developing film may contain disposable negatives which must also be handled in accordance with the TSCA CBI requirements. Giving the facility the option of developing the film may resolve national defense security problems when self-developing film is not satisfactory.

Photographs must be fully documented, following procedures for handling evidentiary materials [Appendix K].

- Arrangement for document availability and copying

The Federal statutes provide broad authorities for document review and copying. If the investigators decide that documents need to be copied, then the investigators should either attempt to use the facility's copier, paying a fee if necessary, or come equipped with a portable copier (Note: renting a portable copier from a local business store often requires advance reservations). A company's refusal to provide documents or refusal to allow copying can be considered similar to a denial of entry [note that the company is NOT required to copy documents for the government without a court order; refusal to copy even with payment is not the same as refusal to provide the documents]. The investigators should note the refusal in log books (including names, titles, and times) and continue with the on-site inspection. At the earliest possible time, supervisors and Federal attorneys should be consulted.

- Company provides overview of facility operations

This provides an opportunity for investigators to learn process operations and to initially identify waste sources.

At the conclusion of the opening conference, information gathering activities begin in earnest. As discussed in preceding inspection strategy section, the next steps may include:

- A general "windshield" tour of the facility
- Split into teams, according to media and process/waste management responsibilities
- Process and laboratory inspections
- Record/document review
- Follow-up interviews

General Facility Tour

The purpose of the general facility tour is to provide investigation team members an "on the ground" orientation and to identify/verify activities requiring further evaluation. The team leader should compile a list of "must see" items, based on the background information review and information obtained during the facility operations overview portion of the opening conference. These could include key process/operations, waste management areas, and areas where suspected violations are occurring. The facility tour (attended by all team members) should include these items; whether facility personnel are provided the specific list depends on whether the company could potentially hide or correct violations. It may be prudent to present the list of specific items in terms of general plant areas to be toured.

The facility tour must be properly structured and knowledgeable facility personnel must accompany the investigators. The route taken may be dictated by facility layout, but material flow should be followed, to the extent possible. The tour should include, as a minimum, raw material storage facilities, manufacturing areas, and waste management units. Team members need to be constantly alert for operations, processes, materials, and waste management activities not previously identified. If a potentially significant operation, unit, or activity (e.g., a waste spill) is observed, "stop the bus" and take a closer look. Any regulatory violations should be properly documented at that time.

The general site tour is also a good time to document conditions with photographs.

Process/Waste Management Evaluation

Once the general site tour is completed, the investigation team may split into smaller units (usually two, sometimes three) that focus on the detailed process evaluations, specific media compliance, or other activities such as laboratory inspection and sample collection. This allows one team to gather process information and begin the compliance evaluation while another team begins concurrent evaluations; all investigators should be watching for potential problems in all media and possible regulatory implications throughout the investigation.

Two key techniques are employed during this part of the investigation: interviews and visual observations. Investigators should employ good interviewing techniques so that the necessary information can be clearly and accurately obtained from facility personnel. Appendix K gives interviewing techniques. Investigators should ask probing questions but never leading questions. Often, the investigator is required to rephrase questions and ask them many times until he/she gets a satisfactory and consistent answer. Body language should also be observed for clues that the facility representative is hedging or that the investigator is starting to key in on a particularly sensitive subject. The investigator should always write down unexpected questions that occur to him/her, especially in situations where these questions cannot be asked promptly. Special care should be taken so that one investigator does not answer another investigator's questions. If the question is posed to a company official, the official should answer even if another investigator knows the answer.

<u>Questioning Facility Personnel</u>

How you ask a question can be more important than the question itself. Try not to give a possible answer when asking a question. For example, the following are poor ways to phrase a question:

- "You don't have any toxic materials around here, do you?"
- "There aren't any buried drums on your property, are there?"
- "You have all the necessary records, don't you?"
- "Your SPCC plan is up to date, isn't it?"

The following questions are better:

- "What kinds of material do you handle?"
- "Are there any materials buried on your property?"
- "Where do you keep the _______ records?"

- "May I see your SPCC plan?" (Then ask an employee about the procedure mentioned in it to verify its application.)

A conclusive question and follow up is often overlooked and taken for granted; however, it is the meat of the inquiry. Without the affirmation of a direct answer to the question at issue, the previous questions were of little value.

Sometimes it is useful to convey the impression that you are there to learn about a facility or its operations and are going to ask a lot of basic questions. This type of "help me learn" attitude will often allow a better line of questioning and more persistence when things become unclear or contradictory. Generally ask open-ended questions first and then clarifying questions as necessary.

Knowledgeable process personnel are usually not used to being interviewed, so it is necessary to operate, initially, in their "comfort zone." They should be asked to describe the process in some detail; the order typically parallels the flow of raw materials and intermediate products toward subsequent processes and the final product(s). During these discussions, waste streams and respective management procedures and related control equipment should be identified. Clarifying questions should focus on raw material/intermediate movements and waste streams produced.

Specifically, information should be obtained on where/how waste is produced, production rates and cycles, spillage or other emissions, housekeeping, floor drains/outlets, waste products, waste minimization, waste mixing/dilution, recent or anticipated modifications, etc. Areas of waste management, treatment, and disposal should also be addressed. Major items of interest include waste spillage/leaks/discharges, how the facility differentiates regulated waste from unregulated waste, physical condition of pollution control equipment, units out of service, operation and maintenance issues, diversions, bypasses and overflows, emergency response capabilities, safety, secondary containment, overloads, waste residuals management, and self-monitoring procedures.

Questions should also be asked about the environmental management program at the facility. Process personnel should explain how they become aware of environmental regulatory requirements and what support they get in taking required compliance actions. Probing follow-up questions may be asked to determine corporate policy toward regulatory compliance. Documenting recalcitrant behavior may have major ramifications in Agency follow-up actions.

Process inspections are then conducted to verify the information presented and discover/discuss "missing" details. These may be done after each process or group of processes is described. Inspection of waste management units may be interspersed with process inspections, depending on their location and facility complexity; a primary consideration is the logical flow of information. Time must also be scheduled to review and copy relevant records, and then for additional interviews to answer questions about the records.

Document Reviews

Before the field investigation begins, each team member should know which reports/records he/she will be responsible for reviewing. These may include inspection logs, annual documents, operating reports, self-monitoring procedures and data, spill clean-up reports, manifests, notifications/certifications, emergency response plans, training records, etc. However, some on-the-spot decisions may have to be made in situations where unexpected information becomes available. The investigator should not limit review to documents specified during the notification or opening conference.

The document review should include determining whether (1) facility personnel have prepared and maintained the required documents, (2) the documents contain all necessary information, (3) the documents have been prepared on time, (4) the documents have been distributed to all necessary parties, and (5) document information is consistent by cross-checking information recorded on more than one document. Document reviews should be done systematically. The investigator should always plan to make copies of documents that cannot be reviewed on-site, or are vitally important in documenting or describing a potential violation. The use of a portable copier should be considered to expedite this procedure.

<u>Sampling</u>

Sampling and sample analysis may be necessary to document noncompliance. Normally, grab samples that are representative and collected by acceptable methods will suffice unless a permit or other legally enforceable document specifies a composite sample (Appendix M presents an example of a sampling guideline to be used in conjunction with detailed SOPs). Samples taken to identify noncompliance with permit requirements must be collected and analyzed consistent with facility permit requirements. Sampling should be considered when the investigator feels that sampling would strengthen a potential enforcement case or help document a potential violation or establish that a facility is subject to regulation. Some situations that may require sampling are:

- Sampling requested by program office (e.g., CWA/NPDES Compliance Sampling Inspection, RCRA Compliance Monitoring Evaluation)

- Leaking drums, tanks, transformers, other containers holding hazardous waste, other toxic materials, or other unexpected or improper releases to the environment

- Unknown waste is found

- Facility's waste analysis data is questionable

- Potential waste misclassification problem is suspected

- Suspicious looking stains or discoloration in waste production/ management areas are unexplained

- Unpermitted discharges are found

- Permitted discharges have a particularly bad appearance or need to be characterized for toxicity/compliance

- Stormwater runoff is suspected of being contaminated

- Receiving waters/sediments are likely to contain toxic/hazardous pollutants

- Contaminated sludges or other waste residuals are being improperly disposed

Closing Conference

The primary purpose of the closing conference is to provide an opportunity for the investigators to discuss __preliminary__ findings with facility representatives, including any potential violations or problems that are uncovered during the investigation. Judgment must be exercised in deciding what findings are presented and how they are represented to facility personnel. Nearly any finding can be discussed if it is presented in the right context. However, the less certain the team leader is about a specific violation or issue, the more reason not to discuss it at the closing conference. In any case, the investigators must clearly state that information provided during the closing conference is preliminary and may change, as a result of additional review. Pollution prevention strategies can also be discussed.

Another important purpose of the closing conference is to resolve any outstanding questions or issues and verify information. Questions or outstanding information requests that cannot be resolved in the closing conference should be compiled into a written, agreed upon document, which is provided to facility representatives before the investigation team leaves the facility, if possible. Each question or information request should be uniquely numbered in this document. Subsequently provided responses should be referenced to the specific request.

Some additional paperwork may need to be completed during the closing conference. This would include signing appropriate inspection forms such as receipt for samples or documents received and declaration of TSCA CBI or the issuance of field citations. Multi-media investigators must be CBI-cleared before they accept any company TSCA CBI information.

Finally, the team leader should be prepared to discuss with facility personnel how and when a copy of the final inspection report can be obtained (e.g., a FOIA request, etc.).

MEDIA SPECIFIC INVESTIGATION PROCEDURES

Presented here, are many of the significant tasks that must be included in each of the media specific inspections. Sample collection and inspection checklists are addressed elsewhere in this manual (Appendix I lists media specific checklists). Media discussed include hazardous waste, water, air, drinking water, toxic substances, pesticides, as well as emergency planning/community right-to-know and Superfund issues. General information on each Act is covered in Appendix A.

RESOURCE CONSERVATION AND RECOVERY ACT (RCRA)

Basic Program

The Resource Conservation and Recovery Act (RCRA) of 1976 is the primary statute regulating the management and disposal of municipal and industrial solid and hazardous wastes. In 1984, RCRA was amended by the Hazardous and Solid Waste Amendments (HSWA) and in 1988 by the Medical Waste Tracking Act (Subpart J of RCRA). The principal objectives of RCRA, as amended, are:

- Promoting the protection of human health and the environment from potential adverse effects of improper solid and hazardous waste management

- Conserving material and energy resources through waste recycling and recovery

- Reducing or eliminating the generation of hazardous waste as expeditiously as possible

The RCRA program consists of four waste management sub-programs designed to meet the Congressional objectives: (1) Subtitle D - Solid Wastes, (2) Subtitle C - Hazardous Wastes, (3) Subtitle I -

Underground Storage Tanks (UST), and (4) Subtitle J - Medical Wastes. This section discusses evaluating compliance under Subtitles C, I, and J.*

<u>Subtitle C, Hazardous Wastes</u>

<u>Evaluating Compliance</u>

Under Subtitle C, hazardous wastes are subject to extensive regulations on generation, transportation, storage, treatment, and disposal. A manifest system tracks shipments of hazardous wastes from the generator until ultimate disposal. This "cradle to grave" management is implemented through regulations and permits.

The investigator must clearly identify investigation objectives, the RCRA regulatory authority (or authorities) with jurisdiction, and establish the facility status under RCRA. RCRA investigations may be performed for several reasons, including:

- Assessing RCRA compliance with regulations and permits
- Reviewing compliance status with respect to an administrative enforcement action
- Reviewing compliance with deadlines in a facility permit
- Responding to alleged violations and/or complaints
- Supporting case development

The regulatory agencies with RCRA authority may be EPA, a designated State agency with full or partial authority, local agencies working with the State, or a combination of the three.

In determining the facility status under RCRA, the investigator must decide whether the facility is a generator, transporter, and/or

---

* *The waste management programs are presented here out of alphabetical sequence because Subtitle D contains the definition of "solid waste" which is helpful in understanding hazardous wastes in Subtitle C. Hazardous wastes are a subset of solid wastes. Subtitle C hazardous wastes are defined specifically in Title 40 of the Code of Federal Regulations (CFR), Part 261.*

treatment, storage, and disposal facility (TSDF), and whether the facility is permitted or has interim status. Generally, EPA Regional and State offices maintain files for the facility to be inspected. Information may include:

- A summary of names, titles, locations, and phone numbers of responsible persons involved in the hazardous waste program

- A list of wastes that are treated, stored, and disposed and how each is managed (for TSDFs)

- A list of wastes generated, their origins, and accumulation areas (for generators)

- Biennial, annual, or other reports required by RCRA and submitted to the regulatory agencies; these include any required monitoring reports

- A detailed map or plot plan showing the facility layout and location(s) of waste management areas

- The facility RCRA Notification Form (Form 8700-12)

- The RCRA Part A Permit Application (for TSDFs)

- The RCRA Part B Permit Application (for TSDFs, if applicable)

- The RCRA permit (for TSDFs, if applicable)

- Notifications and/or certifications for land disposal restrictions (for generators)

Generators

Hazardous waste generators are regulated under 40 CFR Parts 262 and 268. These regulations contain requirements for:

- Obtaining an EPA Identification number
- Determining whether a waste is hazardous
- Managing wastes before shipment
- Accumulating and storing hazardous wastes
- Manifesting waste shipments
- Recordkeeping and Reporting
- Restricting wastes from land disposal (also regulated under Part 268)

The generator regulations vary, depending upon the volume of hazardous wastes generated. The investigator must determine which regulations apply. Additionally, the investigator should do the following:

- Verify that the generator has an EPA Identification Number which is used on all required documentation (e.g. reports, manifests, etc.)

- Confirm that the volume of hazardous wastes generated is consistent with reported volumes. Examine the processes generating the wastes to show that all generated hazardous wastes have been identified. Look for improper mixing or dilution.

- Ascertain how the generator determines/documents that a waste is hazardous. Check to see wastes are properly classified. Collect samples, if necessary.

- Determine whether pre-transport requirements are satisfied, including those for packaging, container condition, labeling and marking, and placarding.

- Determine the length of time that hazardous wastes are being stored or accumulated. Storage or accumulation for more than 90 days requires a permit. Generators storing for less than 90 days must comply with requirements outlined in 40 CFR 262.34.

- Verify RCRA reports and supporting documentation for accuracy, including inspection logs, biennial reports, exception reports, and manifests (with land disposal restriction notifications/certifications).

- Watch for accumulation areas which are in use but have not been identified by the generator. Note: Some authorized State regulations do not have provisions for "satellite storage" accumulation areas.

- Determine whether a generator has the required contingency plan and emergency procedures, whether the plan is complete, and if the generator follows the plan/procedures.

- Determine whether hazardous waste storage areas comply with applicable requirements.

<u>Transporters</u>

Hazardous waste transporters (e.g., by truck, ship, or rail) are regulated under 40 CFR Part 263, which contains requirements for:

- Obtaining an EPA identification number
- Manifesting hazardous waste shipments
- Recordkeeping and reporting
- Sending bulk shipments (by water, rail)

Storage regulations apply if accumulation times at transfer stations are exceeded. Transporters importing hazardous wastes, or mixing hazardous wastes of different Department of Transportation (DOT) shipping descriptions in the same container, are classified as generators and must comply with 40 CFR Parts 262 and 268. Investigators evaluating transporter compliance should do the following:

- Verify that the transporter has an EPA identification number which is used on all required documentation (e.g., manifests)

- Determine whether hazardous waste containers stored at a transfer facility meet DOT pre-transport requirements

- Determine how long containers have been stored at a transfer facility. Storage over 10 days makes the transporter subject to storage requirements

- Verify whether the transporter is maintaining recordkeeping and reporting documents, including manifests, shipping papers (as required), and discharge reports. All required documents should be both present and complete

Treatment, Storage, and Disposal Facilities

Permitted and interim status TSDFs are regulated under 40 CFR Parts 264 and 265, respectively. [Part 264 applies only if the facility has a RCRA permit (i.e., a permitted facility); Part 265 applies if the facility does not have a RCRA permit (i.e., an interim status facility)]. These requirements include three categories of regulations consisting of administrative requirements, general standards, and specific standards (see Table on following page). The investigator should do the following items to determine compliance with Subparts A through E:

- Verify that the TSDF has an EPA identification number which is used on all required documentation.

- Determine what hazardous wastes are accepted at the facility, how they are verified and how they are managed.

- Compare wastes managed at the facility with those listed in the Hazardous Waste Activity Notification (Form 8700-12); the Parts A and B permit applications; and any revisions, and/or the permit.

Table

PERMITTED vs. INTERIM STATUS
RCRA REGULATORY REQUIREMENTS

	Permitted Facility Regulations 40 CFR Part 264	Interim Status Regulations 40 CFR Part 265

Administrative/
nontechnical
requirements
(Subparts A-E)
	• Applicability
	• Facility standards
	  - Applicability
	  - EPA identification number
	  - Required notices
	  - General waste analysis
	  - Security
	  - General inspection requirements
	  - Personnel training
	  - General requirements for ignitable, reactive, or incompatible wastes
	  - Location standards
	• Preparedness and prevention
	• Contingency plan and emergency procedures
	• Manifests and recordkeeping
		• Applicability
		• Facility standards
		  - Applicability
		  - EPA identification number
		  - Required notices
		  - General waste analysis
		  - Security
		  - General inspection requirements
		  - Personnel training
		  - General requirements for ignitable, reactive, or incompatible wastes
		  - Location standards
		• Preparedness and prevention
		• Contingency plan and emergency procedures
		• Manifests and recordkeeping

General standards (Subparts F-H)	• Release from solid waste management units • Closure/post closure requirements	• Groundwater monitoring requirements • Closure/post-closure requirements
Specific standards (Subparts I-R) (Part 264, Subparts I-O, X, AA, and BB) (Part 265, Subparts I-R, AA, BB)	• Containers • Tanks • Surface impoundments • Waste piles • Land treatment • Landfills • Incinerators • Miscellaneous units • Air emission standards for process vents • Air emission standards for equipment leaks	• Containers • Tanks • Surface impoundments • Waste piles • Land treatment • Landfills • Incinerators • Thermal treatment • Chemical, physical, and biological treatment • Underground injection • Miscellaneous units • Air emission standards for process vents • Air emission standards for equipment leaks

- Verify that the TSDF has and is following a waste analysis plan kept at the facility; inspect the plan contents.

- Identify and inspect security measures and equipment.

- Review inspection logs to ensure they are present and complete. Note problems and corrective measures.

- Review training documentation to ascertain that required training has been given to employees.

- Inspect waste management areas to determine whether reactive, ignitable, and incompatible wastes are handled pursuant to requirements.

- Review preparedness and prevention practices and inspect related equipment.

- Review contingency plans; examine emergency equipment and documented arrangements with local authorities.

- Examine the waste tracking system and associated record-keeping/reporting requirements. Required documentation includes manifests and biennial reports, and may include unmanifested waste reports, and spill/release reports. Relevant documents may include on-site waste tracking forms.

- Verify that the operating record is complete according to 40 CFR 264.73 or 265.73.

The investigator can determine compliance with standards in Subparts F through H by doing the following items:

- For permitted facilities, verify compliance with permit standards with respect to groundwater monitoring, releases from solid waste management units, closure/post-closure, and financial requirements (Part 264).

- For interim status facilities required to monitor groundwater, determine what kind of monitoring program applies.

- Depending on the type of investigation, examine the following items to determine compliance:

    - Characterization of site hydrogeology
    - Sampling and analytical records
    - Statistical methods used to compare analytical data
    - Analytical methods
    - Compliance with reporting requirements and schedules
    - Sampling and analysis plan (for content, completeness, and if it is being followed)
    - Condition, maintenance, and operation of monitoring equipment, including wellheads, field instruments, and sampling materials
    - Construction/design of monitoring system
    - Assessment monitoring outline and/or plan
    - Corrective action plan (permitted facilities)

- For waste management units that undergo closure, review the closure plan (including amendments and modifications), plan approval, closure schedule, and facility and regulatory certifications. Examine response actions to any release of hazardous waste constituents from a closed or closing regulated unit.

- For waste management units in post closure care, inspect security measures, groundwater monitoring and reporting, and the maintenance and monitoring of waste containment systems.

- Verify that the owner/operator has demonstrated financial assurance regarding closure.

The technical standards in Part 264 (Subparts I through O and X) and Part 265 (Subparts I through R) govern specific hazardous waste management units used for storage, disposal, or treatment (e.g., tanks, landfills, incinerators). Standards for chemical, physical, and biological treatment at permitted facilities under Part 264 have been incorporated under Miscellaneous Units, Subpart X.* The investigator should do the following:

- Identify all hazardous waste management areas and the activity at each; compare the areas identified in the field with those listed in the permit or permit application, as appropriate. Investigate disparities between actual practice and the information submitted to regulatory agencies.

- Verify that the owner/operator is complying with applicable design, installation, and integrity standards; field-check the design, condition, and operation of waste management areas and equipment.

- Determine how incompatible wastes and ignitable or reactive wastes are managed.

- Verify that the owner/operator is conducting self-inspections where and when required; determine what the inspections include.

- Identify and inspect required containment facilities for condition and capacity; identify leak detection facilities.

---

* *The regulations governing miscellaneous units are intended to address technologies that were difficult to fit into the framework of prior regulations. Miscellaneous units, defined in 40 CFR 260.10, include but are not limited to: open burning/detonation areas, thermal treatment units, deactivated missile silos, and geologic repositories.*

- Determine whether hazardous waste releases have occurred and how the owner/operator responds to leaks and spills.

- Verify that the owner/operator is complying with additional waste analysis and trial test requirements, where applicable.

- Check the closure/post-closure procedures for specific waste management units (e.g., surface impoundments, waste piles, etc.) for regulatory compliance.

- For landfills, determine how the owner/operator manages bulk and contained liquids.

- Field-check security and access to waste management units.

- Determine what are the facility monitoring requirements (for air emissions, groundwater, leak detection, instrumentation, equipment, etc.) and inspect monitoring facilities and records.

When inspecting land treatment facilities, the investigator should also review the following items:

- Soil monitoring methods and analytical data

- Comparisons between soil monitoring data and background concentrations of constituents in untreated soils to detect migration of hazardous wastes

- Waste analyses done to determine toxicity, the concentrations of hazardous waste constituents, and, if food-chain crops are grown on the land, the concentrations of arsenic, cadmium, lead, and mercury. The concentrations must be such that hazardous waste constituents can be degraded, transformed, or immobilized by treatment

- Run-on and run-off management systems

When evaluating compliance of interim status incinerator facilities, the investigator also should review and/or inspect the following items:

- Waste analyses done to enable the owner/operator to establish steady state operating conditions and to determine the pollutant which might be emitted

- General procedures for operating the incinerator during start-up and shut-down

- Operation of equipment monitoring combustion and emissions control, monitoring schedules, and data output

- The incinerator and associated equipment

For permitted incinerators, the investigator must evaluate the incinerator operation against specific permit requirements for waste analysis, performance standards, operating requirements, monitoring, and inspections. The investigator also should do the following:

- Verify that the incinerator burns only wastes specified in the permit

- Verify methods to control fugitive emissions

- Determine waste management practices for burn residue and ash

The investigator evaluating compliance of thermal treatment facilities in interim status also should review the following items:

- General operating requirements, to verify whether steady state operating conditions are achieved, as required

- Waste analysis records, to ensure that (a) the wastes are suitable for thermal treatment, and (b) the required analyses in Part 265.375 have been performed

Thermal treatment facilities permitted under 40 CFR Part 264 Subpart X will have specific permit requirements.

The investigator evaluating compliance of chemical, physical, and biological treatment facilities in interim status also should do the following:

- Determine the general operating procedures.

- Review the waste analysis records and methods to determine if the procedures are sufficient to comply with 40 CFR 265.13.

- Review trial treatment test methods and records to determine if the selected treatment method is appropriate for the particular waste.

- Examine procedures for treating ignitable, reactive, and incompatible wastes for compliance with Subpart Q requirements.

Chemical, physical, and biological treatment facilities permitted under Subpart X will have specific permit requirements.

Owners/operators of TSDFs must also comply with air emission standards contained in Subparts AA and BB of 40 CFR Parts 264 and 265. These subparts establish standards for equipment containing or contacting hazardous wastes with organic concentrations of at least 10%. This equipment includes:

- Process vents
- Pumps in light liquid service
- Compressors
- Sampling connecting systems

- Open-ended valves or lines
- Valves in gas/vapor service or in light liquid service
- Pumps and valves in heavy liquid service, pressure relief devices in light liquid or heavy liquid service, and flanges and other connections

Total organic emissions from process vents must be reduced below 1.4 kg/hr and 2.8 Mg/yr. The other equipment types above must be marked and monitored routinely to detect leaks. Repairs must be initiated within 15 days of discovering the leak.

The facility operating record should contain information documenting compliance with the air emission standards. A complete list of required information is in 40 CFR 264.1035, 264.1064, 265.1035, and 265.1064. Permitted facilities must submit semiannual reports to the Regional Administrator outlining which valves and compressors were not fixed during the preceding 6 months. The investigator can do the following things:

- Visually inspect the equipment for marking.
- Review documentation in the operating record and cross-check this information with that submitted to the Regional Administrator in semiannual reports.

Land Disposal Restrictions

Land disposal restrictions (LDR) in 40 CFR Part 268 are phased regulations prohibiting land disposal* of hazardous wastes unless the waste meets applicable treatment standards [Appendix N].** The

---

*   *Land disposal includes, but is not limited to, placement in a landfill, surface impoundment, waste pile, injection well, land treatment facility, salt dome formation, salt bed formation, underground mine or cave, or placement in a concrete vault or bunker intended for disposal purposes.*

** *Treatment standards are in 40 CFR 268.40 through 43.*

treatment standards are expressed as (1) contaminant concentrations in the extract or total waste, or (2) specified technologies.

Notifications and certifications comprise the majority of required LDR documentation. Notifications tell the treatment or storage facility the appropriate treatment standards and any prohibition levels (California List wastes) that apply to the waste. Certifications are signed statements telling the treatment or storage facility that the waste already meets the applicable treatment standards and prohibition levels.

The regulations divide hazardous wastes into restricted waste groups and apply a compliance schedule of different effective dates for each group (40 CFR 268, Appendix VII).

Investigators evaluating hazardous waste generators for LDR compliance should do the following:

- Determine whether the generator produces restricted wastes; review how/if the generator determines a waste is restricted.

- Review documentation/data used to support the determination that a waste is restricted, based solely on knowledge.

- Learn how/if a generator determines the waste treatment standards and/or disposal technologies.

- Verify whether the generator satisfies documentation, recordkeeping, notification, certification, packaging, and manifesting requirements.

- Ascertain whether the generator is or might become a TSDF and subject to additional requirements.

- Determine who completes and signs LDR notifications and certifications and where these documents are kept.

- Review the waste analysis plan if the generator is treating a prohibited* waste in tanks or containers.

Investigators evaluating TSDFs should do the following:

- Ensure the TSDF is complying with generator recordkeeping requirements when residues generated from treating restricted wastes are manifested off-site.

- Verify whether the treatment standards have been achieved for particular wastes prior to disposal.

- Review documentation required for storage, treatment, and land disposal; documentation may include waste analyses and results, waste analysis plans, and generator and treatment facility notifications and certifications.

Subtitle I - Underground Storage Tanks (USTs)

Evaluating Compliance

Three basic methods are used to determine compliance in most inspections: (1) Interviews of facility personnel, (2) visual/field observations, and (3) document review. Because the tanks are located underground, visual/field observations have limited application in determining compliance for USTs. The UST program relies heavily on the use of documents to track the status and condition of any particular tank.

Interviews with facility personnel are an important starting point when determining compliance with any environmental regulation. Questions regarding how the facility is handling its UST program will give the inspector insight into the types of violations that may be found. Topics to be covered in the interview include:

---

* *Prohibitions on land disposal are explained in 40 CFR 268, Subpart C.*

- Age, quantity, and type of product stored for each tank on-site
- How and when tanks have been closed
- Type of release detection used on each tank (if any); some facilities may have release detection on tanks where it is not required
- Type of corrosion protection and frequency of inspections
- Which tanks have pressurized piping associated with them

Visual/field observations are used to determine if any spills or overfills have occurred that have not been immediately cleaned up. The presence of product around the fill pipe indicates a spill or overfill. Proper release detection methods can also be verified with field observations. During the interviews, ask the facility if monthly inventory control along with annual tightness testing is used. If monthly inventory control is used, check the measuring stick for divisions of 1/8 inch. A field check of the entire facility can also be done to determine if any tanks may have gone unreported. Fillports and vent lines can indicate the existence of a UST.

Documents take up the largest portion of time during a UST inspection. Documents that should be reviewed include:

- Notifications for all UST systems

- Reports of releases including suspected releases, spills and overfills, and confirmed releases

- Initial site characterization and corrective action plans

- Notifications before permanent closure

- Corrosion expert's analysis if corrosion protection is not used

- Documentation of operation of corrosion protection equipment

- Recent compliance with release detection requirements, including daily inventory sheets with the monthly reconciliation

- Results of site investigation conducted at permanent closure.

Document retention rules also apply, so be sure to get all of the documents a facility may be required to keep. To determine if the implementing agency has been notified of all tanks, compare the notifications to general UST lists from the facility. Usually, the facility will keep a list of tanks separate from the notifications and tanks may appear on that list that do not appear on a notification form. Also, compare the notifications to tank lists required in other documents, like the Spill Prevention Control and Countermeasures Plan.

<u>Subtitle J - Medical Wastes</u>

Subtitle J was added to RCRA in November 1988 to address concerns about the management of medical wastes. EPA enacted interim final regulations in March 1989. The regulations, found in 40 CFR Part 259, establish a demonstration program with requirements for medical waste generators, transporters, and treatment, destruction, and disposal facilities (TDDs). The demonstration program is effective in "Covered States" during the period June 22, 1989 to June 22, 1991. The regulations apply to regulated medical waste generated in Connecticut, New Jersey, New York, Rhode Island, and Puerto Rico.

<u>Basic Program</u>

Medical waste is defined in 40 CFR 259.10 as any solid waste generated in the diagnosis, treatment, or immunization of human beings or animals, in related research, biological production, or testing. The following are exempt from 40 CFR Part 259 requirements:

- Any hazardous waste identified or listed under 40 CFR Part 261

- Any household waste defined in 40 CFR 261.4(b)(1)
- Residues from treatment and destruction processes or from the incineration of regulated medical wastes
- Human remains intended to be buried or cremated
- Etiologic agents being shipped pursuant to other Federal regulations
- Samples of regulated medical waste shipped for enforcement purposes

Regulated medical waste is a subset of all medical wastes and includes seven categories:

1. Cultures and stocks of infectious agents
2. Human pathological wastes (e.g., tissues, body parts)
3. Human blood and blood products
4. Sharps (e.g., hypodermic needles and syringes used in animal or human patient care)
5. Certain animal wastes
6. Certain isolation wastes (e.g., waste from patients with highly communicable diseases)
7. Unused sharps (e.g., suture needles, scalpel blades, hypodermic needles)

Etiological agents being transported interstate and samples of regulated medical waste transported off-site by EPA- or State-designated enforcement personnel for enforcement purposes are exempt from the requirements during the enforcement proceedings.

Mixtures of solid waste and regulated medical waste are also subject to the requirements. Mixtures of hazardous and regulated medical waste are subject to the 40 CFR Part 259 requirements only if shipment of such a mixture is not subject to hazardous waste manifesting (e.g., the hazardous waste is shipped by a conditionally exempt generator).

Generators, transporters, and owners or operators of intermediate handling facilities or destination facilities who transport, offer for transport, or otherwise manage regulated medical waste generated in a Covered State must comply with the regulations even if such transport or management occurs in a non-Covered State. Vessels at port in a Covered State are subject to the requirements for those regulated medical wastes transported ashore in the Covered State. The owner or operator of the vessel and the person(s) removing or accepting waste from the vessel are considered co-generators of the waste.

A generator who either treats and destroys or disposes of regulated medical waste on-site [e.g. incineration, burial, or sewer disposal covered by section 307(b) through (d), of the Clean Water Act] is not subject to tracking requirements for that waste. However, such on-site waste management may subject the generator to additional Federal, State, or local laws and regulations.

<u>Evaluating Compliance</u>

The inspector should evaluate whether the generator has determined what regulated medical waste streams are generated and/or managed. Generators of less than 50 pounds per month are exempt from certain transportation, and tracking requirements. Compliance should also be evaluated by observing the following:

- Prior to shipping waste off-site: Are wastes segregated? Are wastes packed in the appropriate containers? If containers are reused, are they decontaminated? Are containers properly marked?

- Does the generator use tracking forms? Are copies of the forms and any exception reports kept for 3 years? Does the generator export medical waste for treatment, destruction, or disposal? If so, the generator must request that the destination facility provide written confirmation that the waste was received; an exception report must be filed if such a confirmation is not received within 45 days. If the generator incinerates medical waste on-site, are the recordkeeping and reporting regulations for on-site incinerators followed?

The transportation requirements apply to transporters, including generators who transport their own waste, and owners and operators of transfer facilities engaged in transporting regulated medical waste generated in a Covered State. The inspector should verify that:

- The proper labeling and marking of regulated medical waste accepted for transportation has taken place or has been done.

- If the waste is handled by more than one transporter, did each transporter attach a water resistant identification tag below the generator's marking? Is the required information on the tag?

- The transporter submitted the required notification(s) for each Covered State.

- The vehicles are fully enclosed, leakproof, maintained in sanitary condition, secured when unattended, and marked with the proper identification.

- The applicable requirements for rail shipments are followed.

- Tracking forms are used properly.

- Recordkeeping and reporting requirements are followed.

The requirements for treatment, destruction, and disposal facilities apply to owners and operators of facilities that receive regulated medical waste generated in a Covered State, including facilities located in non-Covered States that receive regulated medical waste generated in a Covered State. The facilities include destination facilities, intermediate handlers, and generators who receive regulated medical waste required to be accompanied by a tracking form. The inspector should verify the following:

- Are tracking forms used and properly completed?

- Are tracking form discrepancies resolved?

- Are the recordkeeping requirements followed?

- Is any additional information required by the Administrator reported?

For rail shipments of regulated medical waste, the inspector should determine whether the tracking forms are used properly.

Pollution Prevention

EPA is developing an Agency-wide policy for pollution prevention. Present authorities were established in the 1984 Hazardous and Solid Waste Amendments to RCRA [Section 3002]. The October 1990 Pollution Prevention Act established pollution prevention as a national priority.

Evaluating Compliance

EPA has developed a policy regarding the role of inspectors in promoting waste minimization (OSWER directory number 9938.10). As stated in the policy, to evaluate compliance, the inspector should:

- Check hazardous waste manifests for a correctly worded and signed waste minimization certification.

- Determine whether this certification was manually signed by the generator or authorized representative.

- Confirm that a waste minimization program is in place by requesting to see a written waste minimization plan, or requesting that the plan be described orally, or requesting that evidence of a waste minimization program be demonstrated.

The inspector can, and should visually check for evidence of a "program in place" on-site.

- Check the Biennial Report and/or Operating Record of generators and TSDs, as appropriate. These documents are to contain descriptions of waste minimization progress and a certification statement. If known omissions, falsifications, or misrepresentations on any report or certification are suspected, criminal penalties may apply and the case should be referred for criminal investigation.

- Check any waste minimization language included in the facility's permits, any enforcement order, and settlement agreements. Verify that any waste minimization requirements are being satisfied.

The policy also states that the inspector should promote waste minimization by:

- Being familiar with recommending, and distributing waste minimization literature.

- Referring the facility to the appropriate technical assistance program for more specific or technical information.

- Providing limited, basic advice to the facility of obvious ways they can minimize their waste. This advice should be issued in an informal manner with the caveat that it is not binding in any way and is not related to regulatory compliance.

The multi-media inspection team can also document cross-media transfers of waste streams, which can result in false claims of waste minimization. For example, a facility could treat a solvent wastewater stream in an air stripper that has no air pollution control devices. On paper, the amount of solvent discharged to a land disposal unit or sewer system could show a reduction, but the pollutants are going into the air, possibly without a permit. Another example would be a facility claiming a reduction in hazardous waste generated because the waste stream was delisted.

<u>CLEAN WATER ACT (CWA)</u>

EPA establishes national water quality goals under the CWA. Water pollution from industrial and municipal facilities is controlled primarily through permits limiting discharges. Permit limits are based on effluent guidelines for specific pollutants, performance requirements for new sources, and/or water quality limits. Permits also set schedules and timetables for construction and installation of needed equipment. Sources which discharge indirectly to a municipal treatment plant are subject to pretreatment standards. Other key provisions of the CWA require permits for discharge of dredged and fill materials into waters (including wetlands) and requirements for reporting and cleaning up spills of oil or hazardous material. Nonpoint sources of water pollution, such as runoff from agricultural fields, are addressed through programs to implement Best Management Practices.

Although the investigator(s) responsible for determining facility compliance with Clean Water Act requirements should focus on issues identified below, they should be aware of the inter-relationship with other laws, regulations, etc. For example, sludge generated at a wastewater treatment plant (WWTP) may be regulated under solid waste disposal laws (Toxicity Characteristic) and substances used/generated at the WWTP may be subject to reporting requirements (EPCRA reporting for chlorine).

<u>Basic Program</u>

Wastewater compliance components can be generally categorized into the following groups:

- Control and treatment systems
- Self-monitoring systems (including both field and laboratory measurements)
- Operation and maintenance (O&M)

- Best Management Practices (BMP)
- Spill Prevention Control and Countermeasure (SPCC) Plan

Before the inspection, the investigators should determine the "yardstick" by which facility compliance will be measured. To do so, the investigator must obtain and review copies of the discharge permit, permit application, discharge monitoring reports (DMRs), treatment facility plot plans, and any additional required plans (SPCC, etc.).

<u>Evaluating Compliance</u>

<u>Control and Treatment Systems</u>

Wastewater control and treatment systems should be evaluated for adequacy and compliance with permit or other requirements (consent decrees, etc.) through record review and on-site inspection. This includes, but is not limited to, the following:

- Determine if all wastewaters generated by the facility are adequately controlled, recycled, directed to the wastewater treatment plant (on or off-site), discharged through an outfall regulated by a National Pollutant Discharge Elimination System (NPDES) permit, etc.

- Identify any wastewater discharges directly to a receiving waterbody that are not included in a facility NPDES Permit.

- For off-site wastewater treatment, determine if the discharge is required to meet pretreatment standards. Review any applicable standards and appropriate wastewater characterization data, as necessary. Pretreatment checklists are available in some Regional offices.

- For on-site wastewater treatment, determine if the wastewater treatment plant has the appropriate unit processes and is properly sized to effectively treat the quality and quantity of wastewater generated by the facility.

- Review operations records and DMRs to determine if the facility has exceeded its NPDES permit limits.

<u>Self-monitoring Systems</u>

Self-monitoring consists of flow and water quality measurements and sampling by the facility in addition to laboratory analyses of water samples required by the NPDES permit program. The NPDES/ pretreatment permits normally identify self-monitoring requirements. There are usually two components to the self-monitoring system evaluation, as discussed below:

Field - Confirm that acceptable sampling and flow measurements, as specified by the NPDES/pretreatment permits, are conducted at the correct locations, with the proper frequency, and by acceptable equipment and methods. Determine if all necessary calibrations and O&M are performed. Approved procedures are to be used in the collecting, preserving, and transporting of samples [40 CFR 136.3(e)].

Laboratory - Evaluate laboratory procedures affecting final reported results including:

- Sample preservation methods and holding times
- Chain-of-custody
- Use of approved procedures (40 CFR 136 or approved alternatives)
- Adequacy of personnel, equipment, and other components of laboratory operations
- Adequacy of quality assurance/quality control program
- Recordkeeping and calculations

Evaluate how the data are entered into laboratory notebooks or computers; sign-off procedures used; analysis of spikes, blanks, and reference samples; how the lab data are transposed onto the official, self-monitoring report forms (DMRs) sent to the regulatory agency; and the extent and capability of outside contract laboratories, if used.

Operation and Maintenance

Most NPDES discharge permits have standard language that requires proper facility operation and maintenance [40 CFR 122.41(e)]. The investigator should:

- Determine if wastewater treatment processes are operated properly through visual inspection and records review.

- Observe the presence of solids, scum, grease, and floating oils or suspended materials (pinpoint floc, etc.), odors, and weed growth in the treatment units. Note appearance of wastewater in all units.

- Identify all out-of-service processes and determine cause.

- Determine level of maintenance by observing condition of equipment (pumps, basins, etc.) and reviewing records (outstanding work orders, spare parts inventories).

- Identify handling, treatment, and disposal of sludges and other residues generated from processes and wastewater treatment system.

Best Management Practices (BMP) Plan

Determine if the facility handles toxic materials and if a BMP plan is required (40 CFR 125, Subpart K or by NPDES permit). If applicable, review BMP Plan or BMP Permit requirements. Determine if facility is following required provisions. Review any records required by the plan for adequacy.

SPCC Plan

Determine if the facility is required to develop and implement an SPCC Plan (40 CFR 112) for storage/handling and spill control of specified substances. A facility is required to have an SPCC plan if it stores oil and/or oil products and:

- Underground capacity exceeds 42,000 gallons
- Aboveground storage capacity exceeds 1,320 gallons
- Any single aboveground container exceeds 660 gallons
- A spill could conceivably reach a water of the United States

Obtain a copy of the plan and required records to assess compliance with the plan provisions. The plan should be certified by a registered professional engineer with approval and implementation certified by the proper facility official. Identify and visually inspect all regulated tanks and equipment including containment and run-off control systems and procedures. Investigate any evidence of spilled materials. Discuss training and associated procedures with facility personnel. Review applicable records (spill reports, tank and piping inspection reports, and loading/unloading equipment inspection reports).

CLEAN AIR ACT (CAA)

The Clean Air Act is the legislative basis for air pollution control regulations. It was first enacted in 1955 and later amended in 1963, 1965, 1970, 1977, and 1990. The 1955 Act and the 1963 Amendments called for the abatement of air pollution through voluntary measures. The 1965 amendments gave Federal regulators the authority to establish automobile emission standards.

Basic Program

The Clean Air Act Amendments of 1970 significantly broadened the scope of the Act, forming the basis for Federal and State air pollution control regulations. Section 109 of the 1970 Amendments called for the attainment of national ambient air quality standards (NAAQS, 40 CFR 50) to protect public health and welfare from the known or anticipated adverse effects of six air pollutants (as of 1990 the standards were for small particulates, sulfur dioxide, nitrogen dioxide, carbon monoxide, ozone, and lead). The states were required to develop and submit to EPA implementation plans that were designed to achieve the NAAQS. These state implementation plans (SIPs) contained regulations that limited air emissions from stationary and mobile sources. They were developed and submitted to EPA on a continuing basis and became federally enforceable when approved.

Section 111 of the 1970 Amendments directed EPA to develop standards of performance for new stationary sources. These regulations, known as New Source Performance Standards (NSPS, 40 CFR 60), limited air emissions from subject new sources. The standards are pollutant and source specific. Appendix L contains a list of the NSPS sources as of July 1, 1990 and also lists sources subject to NSPS continuous emission monitoring (CEM) and continuous opacity monitoring (COM) requirements.

Section 112 of the 1970 Amendments directed EPA to develop standards for hazardous air pollutants. These regulations, known as the National Emission Standards for Hazardous Air Pollutants (NESHAPs, 40 CFR 61), limited hazardous air emissions from both new and existing sources. (Appendix L) contains a list of sources subject to NESHAPs as of July 1, 1990. These standards are incorporated into the SIPs, usually by reference to the EPA standard.

The Clean Air Act Amendments of 1977 addressed the failure of the 1970 Amendments to achieve the NAAQS by requiring permits for major new sources. The permit requirements were based on whether the source was located in an area that met the NAAQS (attainment areas, 40 CFR 81) or in an area that did not meet the NAAQS (nonattainment areas). The permit program for sources in attainment areas was referred to as the prevention of significant deterioration (PSD) program.

The Clean Air Act Amendments of 1990 significantly expanded the scope of the Act. Section 112 amendments essentially replaced the NESHAPs with a new program called "Title III - Hazardous Air Pollutants." Title III listed 189 hazardous air pollutants [Appendix O] and required EPA to start setting standards for categories of sources that emit these pollutants within 2 years (1992) and finish setting all standards within 10 years (2000). It also contains provisions for a prevention-of-accidental-releases program.

Title V of the 1990 Amendments requires EPA to promulgate a permitting program that will be implemented by the states no later than November 15, 1994. The permits will include enforceable emission standards, and reporting, inspection, and monitoring requirements

Title VII of the 1990 Amendments gives EPA enhanced enforcement authority. The Agency may initiate enforcement proceedings for SIP and permit violations if the state does not take enforcement action. Title VII also provides for criminal penalties for Clean Air Act violations.

<u>Evaluating Compliance</u>

The following procedures are used to evaluate compliance with the Clean Air Act.

Before an on-site inspection the documents listed below should be obtained from state or EPA files and reviewed to determine what regulations apply and what compliance problems may exist.

- The state air pollution control regulations contained in the SIP (State regulations and permits form the basis for the air compliance inspection and will vary from state to state.)

- The state operating and construction permits

- The most current emissions inventory (check for sources subject to SIP, NSPS, and NESHAPs requirements)

- The volatile organic compound (VOC) emissions inventory (The VOC inventory may not be included in the emissions inventory but reported separately under Title III Form R submittals. See Emergency Planning and Community Right-to-Know section.)

- The consent decrees/orders/agreements still in effect and related correspondence

- The most recent inspection reports

- The most recent monthly or quarterly CEM/COM reports

- AIRS Facility Subsystem (AFS) reports

- Process descriptions, flow diagrams, and control equipment for air emission sources

- Facility plot plan that identifies and locates the air pollution emission points

The on-site inspection should include a review of the records and documents listed below:

- Process operating and monitoring records to determine if permit requirements are being followed

- Fuel analysis reports (including fuel sampling and analysis methods) to determine if sulfur dioxide emission limits and/or other fuel requirements are being met

- Reports of process/control equipment malfunctions causing reportable excess emissions (refer to SIP to determine reportable malfunctions and report requirements)

- Source test reports to determine if NSPS, NESHAPs, and/or major sources have demonstrated compliance with emission standards

- CEM reports to determine if NSPS and SIP reporting requirements are being met (reported emissions should be checked against raw data for accuracy and reported corrective actions should be checked for implementation)

- CEMS/COMS certification tests (relative accuracy and calibration drift) to verify that performance specifications at 40 CFR 60, Appendix B are met

- Records and reports specified in SIP regulations, NSPS and NESHAP subparts, and applicable permits

The on-site inspection should also include the following:

- Visible emission observations (VEOs), by inspectors certified to read smoke within the last 6 months, to determine compliance with SIP, NSPS, or NESHAPs opacity limits (document noncompliance with EPA Method 9, 40 CFR 60, Appendix A)

- A check of real time CEM measurements to determine compliance SIP, NSPS or NESHAPs limits (opacity CEM measurements can be compared against VEOs)

- A review of CEM/COMS calibration procedures and frequency to determine if the zero/span check requirements and analyzer adjustment requirements of 40 CFR 60 are being met

- Observations of process and control equipment operating conditions to determine compliance with permit conditions (if no permit conditions apply, control equipment operating conditions can be compared to baseline conditions from stack tests or manufacturers specifications for proper operation)

- A review of all sources to determine if existing, new, modified or reconstructed sources have construction and operating permits required by SIP (note other process changes that may not require a permit but could effect emissions)

- Observation of control equipment operating conditions and review of equipment maintenance practices and records to determine proper operation of control equipment

SAFE DRINKING WATER ACT (SDWA)

Basic Program

Public drinking water supply systems (i.e., serve at least 25 people) are regulated by the Safe Drinking Water Act (SDWA), as amended. EPA sets standards for the quality of water that can be served by public water systems, [known as Maximum Contaminant Levels (MCLs)]. Public systems must sample their water periodically and report findings to the State (or EPA, if the State has not been delegated the authority to enforce the SDWA). They must notify consumers if they do not meet the standards or have failed to monitor or report. EPA is on a statutory schedule for promulgating a large number of new MCLs.

Evaluating Compliance

Water Supply Systems

The Underground Injection Control (UIC) program was developed pursuant to the Safe Drinking Water Act (SDWA) (Public Law 93-523), Part C - Protection of Underground Sources of Drinking Water (40 CFR Parts 124 and 144 through 148).

The UIC program regulates five classes of injection wells, summarized as follows:

Class I	Industrial, municipal, or hazardous waste disposal beneath the lowermost underground source of drinking water (USDW)
Class II	Oil- and gas-related wells used for produced fluid disposal, enhanced recovery, hydrocarbon storage, etc.
Class III	Mineral extraction wells
Class IV	Hazardous or radioactive waste disposal above or into a USDW
Class V	Injection wells not included in Classes I through IV

Monitoring requirements for water supply systems and whether or not the system can be reasonably expected to routinely provide safe potable water should be determined. Many facilities purchase their potable water supply from a nearby municipality. If no further treatment is provided (e.g., chlorination by the facility), the facility remains a "consumer" rather than becoming a "supplier," and consequently does not have the monitoring or reporting requirements that a supplier would have. Nevertheless, the facility does have a responsibility to assure that their actions do not result in contamination of the municipal water supply (e.g., through cross-connection). The audit team should be alert to these possibilities.

Inspectors should:

- Verify public water system records of monitoring and reports of exceeding MCLs.

- Interview water system personnel to identify potential operations and maintenance problems.

- Obtain water source, treatment, and service area information.

UIC inspector should determine the following:

- Injection well construction
- Potential pathways of endangerment to underground sources of drinking water (USDWs)
- Protection of USDWs from endangerment
- Frequency and type of mechanical integrity testing (MIT)
- Annular pressure
- Annular pressure monitoring
- Radioactive tracer surveys
- Installation methods for well plugging
- Remedial operations
- Applicability of Land Disposal Restrictions to injection well operations
- Recordkeeping and evidence documentation
- Outlets for floor drains
- Connection to "dry" wells

Several states and industries have requested approval of various alternative mechanical integrity testing methods or variances to accommodate special local hydrogeological conditions, historical practices, or industry interests. Inspectors and field investigators should be cautioned to keep current with special permit conditions and the status of any pending approvals/denials of alternative mechanical integrity testing procedures and variances.

TOXIC SUBSTANCES CONTROL ACT (TSCA)

This section describes those specific aspects of toxic chemical control that are addressed by the Toxic Substances Control Act (TSCA) and its associated rules and regulations (40 CFR Parts 702 through 799).

Basic Program

The regulation of toxics under TSCA is subdivided into two components for Agency enforcement program management purposes.

1. "Chemical control" covers enforcement aspects related to specific chemicals regulated under Section 6 of TSCA, such as polychlorinated biphenyls (PCBs), chlorofluorocarbons (CFCs), and asbestos.

2. "Hazard evaluation" refers to the various recordkeeping, reporting, and marketing submittal requirements specified in Sections 5, 8, 12, and 13 of TSCA; although, some elements of what might be termed "chemical control" are also addressed in these sections. Sections 12 and 13 of TSCA, which pertain to chemical exports and imports, respectively, will not be covered in this manual due to their special nature and unique requirements.

Prior to discussing TSCA activities[*] at a facility, the investigator must present appropriate facility personnel with copies of the following two TSCA audit forms [Appendix P]:

1. Notice of Inspection - Shows purpose, nature, and extent of TSCA audit

2. TSCA Inspection Confidentiality Notice - Explains a facility's rights to claim that some or all the information regarding toxic substance handling at the facility is to be considered as TSCA Confidential Business Information (CBI)

Before leaving the site, the following two forms [Appendix P] must be completed, as appropriate.

1. Receipt for Samples and Documents - Itemizes all documents, photos, and samples received by the investigator during the audit

---

[*] *All personnel handling material claimed as Confidential Business Information under TSCA must be cleared for access to that material in accordance with Agency procedures. An annual update is required.*

2.   Declaration of CBI[*] - Itemizes the information that the facility claims to be TSCA CBI

Evaluating Compliance

Chemical Control

Although the controlled substances most frequently encountered during multi-media investigations are polychlorinated biphenyls (PCBs), the investigator should determine if other regulated toxic substances are present at the facility. Currently these include metal working fluids (Part 747), fully halogenated chlorofluoroalkanes (40 CFR 762), and asbestos (40 CFR 763); additional toxic substances may be regulated in the future. Because the probability of finding PCBs and PCB items at a facility is greater than finding other TSCA-regulated substances, the following discussion is directed toward an evaluation of compliance with proper PCB and PCB item handling procedures. Should other TSCA-regulated substances be present, the investigator should consult the regulations for appropriate requirements.

Management of PCBs/PCB items is regulated under 40 CFR 761. In general, these regulations address recordkeeping, marking and labeling, inspections, storage, and disposal.

Facilities which store and/or dispose of PCBs and PCB items should have EPA-issued Letters of Approval which contain facility operating and recordkeeping requirements in addition to those specified in 40 CFR 761. The investigator must obtain a copy of these approvals and any subsequent notifications to evaluate facility compliance. The inspector should review Part 761.30 to identify uses of PCB transformers which are prohibited beginning October 1, 1990, but with effective dates extending to October 1, 1993. The inspector should also review the requirements found in

---

[*]   *These forms are generally completed during the closing conference. During the opening conference, facility personnel should be made aware that the latter form is used to itemize TSCA CBI material.*

Part 761.30 which allow the installation of PCB transformers for emergency use.

In general, the compliance evaluation includes obtaining and reviewing information from Federal, State, and local regulatory agency files; interviewing facility personnel regarding material handling activity; examining facility records and inspecting material handling units. Specific investigation tasks include:

- Inspect all in-service electrical equipment, known or suspected of containing PCBs, for leaks or lack of proper marking. A similar inspection should also be made of any equipment that the facility is storing for reuse. Make certain that any remedial actions were quick and effective in the case of leaks, spills, etc.

- If the above equipment includes any PCB transformers make certain that all relevant prohibitions are being met, such as those involving enhanced electrical protection, as well as other requirements in the Use Authorization section of the PCB Rule. Likewise with large PCB capacitors. Make certain that any hydraulic or heat transfer systems suspected of containing PCB fluids have been properly tested.

- Determine whether the facility is involved with servicing PCB items or using/collecting/producing PCBs in any manner. If so, make certain that the appropriate requirements of the PCB Rule are being met.

- Determine whether the facility is involved with either the storage or disposal of PCBs/PCB items. Inspect all storage for disposal facilities for proper containment, leaking items, proper marking, dates/time limits, location, protection from elements, and other necessary requirements. If the facility disposes of PCBs, make certain that proper methods are being employed and that design and operation of disposal units is in accordance with regulatory requirements.

- Determine whether storage/disposal facilities are complying with the notification and manifesting requirements contained in Subpart K of the PCB Rule.

- Thoroughly review, for purposes of adequacy and regulatory compliance, all records and reports required by the PCB Rule including the following:

  - Annual documents
  - Inspection logs
  - PCB transformer registration letters
  - Manifests/certificates of destruction
  - Test data
  - Spill clean-up reports
  - EPA issued permits or Letters of Approval
  - SPCC plan, if one is necessary
  - Operating records
  - Notification of PCB activity

<u>Hazard Evaluation</u>

Establishing compliance with the various hazard evaluation aspects of TSCA is best accomplished through review and evaluation of the recordkeeping, reporting, and submittal data required by the various regulatory components of Sections 5 and 8. In general, Section 5 addresses new chemicals (i.e., those not on the TSCA Chemical Substances Inventory) and Section 8 provides for control of existing chemicals (i.e., those chemicals that are on the TSCA Chemical Substances Inventory).

Much of the information obtained and reviewed under these two sections of TSCA will be declared "TSCA Confidential Business Information" (CBI) by company officials and, thus, special security procedures must be followed during review and storage of the documents, as discussed elsewhere.

The glossary [Appendix Q] and 40 CFR Parts 703 through 723 should be consulted for an explanation of TSCA terms and definitions. The following list summarizes the different objectives for inspections of the key TSCA Sections 5 and 8 components.

1.   <u>Premanufacture Notification (PMN)</u>

   a.   Verify that all commercially manufactured or imported chemicals are either on the TSCA Chemical Substances Inventory, are covered by an exemption, or are not subject to TSCA.

   b.   Verify that commercial manufacture or import of new chemicals did not begin prior to the end of 90-day review date [Appendix A, page A-24], and not more than 30 days before the Notice of Commencement (NOC) date. If commercial manufacture or import has not begun, verify that no NOC has been submitted .

   c.   Verify the accuracy and documentation of the contents of the PMN itself.

2.   <u>Research and Development (R&D) Exemption</u>

   a.   Verify that the recordkeeping and notification requirements are being met for all R&D chemicals.

   b.   Verify that "Prudent Laboratory Practices" and hazardous data searches are adequately documented.

3. <u>Test Marketing Exemption (TME)</u>

   a. Verify that the conditions spelled out in the TME application are being met, particularly with respect to dates of production, quantity manufactured or imported, number of customers and use(s).

   b. Verify that the TME recordkeeping requirements are being met.

4. <u>Low Volume Exemption (LVE) and Polymer Exemption (PE)</u>

   a. Verify that specific conditions of the exemption application are being met, and that all test data have been submitted.

   b. For an LVE, verify that the 1,000-kg limit per 12-month period has not been exceeded. For a PE, assure that the chemical structure and monomer composition(s) are accurate.

   c. Verify that recordkeeping requirements for both LVEs and PEs are being met.

5. <u>5(e)/5(f) Order, Rule, or Injunction</u>

   a. Verify that all conditions of the order, rule, or injunction are being followed, including use of protective equipment, glove testing, training, and recordkeeping.

   b. If a testing trigger is specified, verify production volume and status of testing activity.

6.   <u>Significant New Use Rule (SNUR)</u>

    a.   Verify that no commercial production has occurred prior to the 90-day review date.

    b.   Verify that SNUR notices have been submitted for all applicable manufactured, imported, or processed chemicals.

    c.   Verify technical accuracy of SNUR submittal and completeness of required recordkeeping.

7.   <u>Bona Fide Submittals</u>

Determine the commercial production (or import) status and R&D history of those bona fide chemicals not found on the confidential 8(b) inventory. Verify findings against applicable PMN, TME, or other exemption.

8.   <u>Section 8(a) Level A PAIR and CAIR Report</u>

    a.   Determine if Preliminary Assessment Information Rule (PAIR) and Comprehensive Assessment Information Rule (CAIR) reports have been submitted for all 8(a) Level A listed chemicals manufactured or imported by the facility.

    b.   Verify the accuracy of submitted PAIR information, particularly the reported figures for total production volume and worker exposure levels.

    c.   Verify the accuracy of submitted CAIR information and if the report meets the date specified in the regulation.

9. Section 8(b) Inventory Update Rule (IUR)

  a. Verify the accuracy of the information submitted in response to the IUR.

  b. Determine that required information was submitted by the prescribed deadline for all chemicals subject to IUR.

10. Section 8(c) Recordkeeping

  a. Determine if the facility has a Section 8(c) file and that allegations of significant health and environmental harm on record are properly filed and recorded.

  b. Determine that all applicable allegations have been recorded and filed.

  c. Determine if the facility has a written Section 8(c) policy and if the policy includes outreach to the employees.

11. Section 8(d) Reporting

Determine if copies (or lists) of all unpublished health effects studies have been submitted by manufacturers, importers, and processors for any Section 8(d) listed chemical.

12. Section 8(e) Reporting

  a. Verify that all Section 8(e) substantial risk reports to the Agency were accurate and submitted within the required time frames.

  b. Verify that all substantial risk incidents and/or test results have been reported to EPA.

c.   Determine that the company has an adequate written policy addressing Section 8(e), and that it relieves employees of individual liability.

FEDERAL INSECTICIDE, FUNGICIDE, AND RODENTICIDE ACT (FIFRA)

Basic Program

Pesticides are regulated by FIFRA and regulations developed under delegated State programs. Under FIFRA, pesticide products must be registered by EPA before they are sold or distributed in commerce. EPA registers pesticides on the basis of data adequate to show that, when used according to label directions, they will not cause unreasonable adverse effects on human health or the environment.

To ensure that previously registered pesticides meet current scientific and regulatory standards, in 1972 Congress amended FIFRA to require the "reregistration" of all existing pesticides.

Evaluating Compliance

The following list is for use in conjunction with specific storage/ use/disposal requirements found on pesticide labels. FIFRA requires a written notice of inspection and written receipt for samples and documents collected.

- Determine types and registration status of all pesticides produced, sold, stored, and used at the facility, particularly if any are restricted or experimental use pesticides.

- Determine use(s) of each pesticide.

- Determine certification status of facility/handlers.

    - Verify who certifies facility/pesticide handlers [EPA, State, Department of Defense (DOD)].

    - Determine if commercial or private application.

    - If restricted-use pesticides are used, check if pesticide applicators are authorized to use these pesticides.

    - Check expiration dates on licenses/certificates.

- Review applicable records.

    - Check previous audit records and complaints.

    - Check application records.

    - Check restricted-use pesticides records (must be kept at least 2 years). Document suspected violations accordingly.

    - Check inventory records.

    - Check training records.

    - Check equipment repair records.

- Inspect storage, mixing/loading, and container disposal areas.

    - Check bulk storage areas for compliance with Federal/ State rules.

- Check location, ventilation, segregation, shelter, and housekeeping of pesticide storage/handling areas. Check security, fire protection, and warning signs, as may be required by State regulations.

- Check mixing equipment/procedures for reducing handlers' exposures to pesticides.

- Check for safety equipment/procedures/use.

- Check container cleanup and disposal procedures.

- Pesticide waste disposal

  - Check to see that pesticides are disposed of in accordance with applicable label and RCRA requirements.

- Determine measures taken to ensure worker safety.

  - Check pesticide use records for re-entry time limit notation.

  - Check pesticide use records for record of informing farmer or warning workers and/or posting fields.

  - Provide farmer and/or applicator copy of current worker protection standards.

- Observe actual pesticide application.

  - Observe mixing/loading and check calculations for proper use dilution.

  - Observe when spray is turned on/off with respect to ends of field.

- Watch for drift or pesticide mist dispersal pattern.

- Note direction of spraying pattern and trimming techniques.

- Record wind speed and direction, air temperature, and relative humidity.

- Observe application with respect to field workers, houses, cars, power lines, and other obstacles.

- Determine if applicator and assisting personnel are wearing safety gear required by the label.

EMERGENCY PLANNING AND COMMUNITY RIGHT-TO-KNOW ACT (EPCRA)

Basic Program

The Emergency Planning and Community Right-to-Know Act (EPCRA) of 1986 is a free-standing law contained within the Superfund Amendments and Reauthorization Act (SARA) of 1986. EPCRA is also commonly known as SARA Title III. EPCRA requires dissemination of information to State and community groups and health professionals on chemicals handled at regulated facilities.

An EPCRA audit verifies that the facility owner/operator has notified State and local agencies of regulated activities; has submitted information to specific State and local agencies; and has prepared and submitted all other required reports.

<u>Evaluating Compliance</u>

<u>Emergency Planning (Sections 301 through 303)</u>

EPA promulgated regulations which identify extremely hazardous substances and the levels to be regulated under EPCRA. The inspector should determine whether the facility is subject to EPCRA regulation. If the facility does meet the requirements, the inspector should verify whether the facility owner/operator:

- Notified the State emergency response agency and the local emergency planning committee that the facility is regulated under EPCRA

- Designated a facility emergency coordinator to assist the local emergency planning committee in the planning process

- Notified the local emergency planning committee of the emergency coordinator's identity

<u>Emergency Notification (Section 304)</u>

The owner/operator of a facility subject to EPCRA must immediately report releases of hazardous substances. Substances subject to this requirement are the extremely hazardous substances listed in 40 CFR Part 355 and substances subject to the emergency notification requirements under CERCLA Section 103(a) or (c). The inspector should verify whether an immediate notification was made to the:

- State emergency response commission
- Local emergency planning committee
- National Response Center for spills involving CERCLA reportable quantities

Community Right-to-Know Requirements (Sections 311 through 312)

Manufacturing facilities subject to the Occupational Safety and Health Act (OSHA) Hazardous Communication Regulation (29 CFR Part 1910) are required to prepare Material Safety Data Sheets (MSDS) for each hazardous chemical handled at the facility. Manufacturing facilities covered are contained within Standard Industrial Classification (SIC) Codes 20 through 39. OSHA revised its Hazardous Communication Regulation, effective September 23, 1987, to require that MSDSs be prepared by nonmanufacturing facilities. The inspector should verify that the facility owner/operator has sent the following to the State emergency response commission, the local emergency planning committee and the local fire department:

- MSDS or a list of chemicals covered by MSDS found at the facility

- An annual inventory of hazardous chemicals found at the facility

Toxic Chemical Release Reporting (Section 313)

Covered facilities (40 CFR Part 372.22) that manufacture, import, process, or use certain chemicals must annually report releases to the environment. The inspector should determine whether the facility owner/ operator is required to submit a report (Form R) by July 1 for the preceding calendar year(s). All the following conditions must apply at the facility in order to meet the reporting requirements:

- The facility has 10 or more full-time employees

- An operation(s) identified in SIC Codes 20 through 39 is present

- The amount of chemical(s) handled exceeds the applicable threshold quantity

COMPREHENSIVE EMERGENCY RESPONSE, COMPENSATION, AND LIABILITY ACT (CERCLA)

The Superfund law of 1980 (CERCLA) including the SARA amendments of 1986 authorizes EPA to clean up hazardous substances at closed and abandoned waste sites and to recover the cost of cleanup and associated damages from the responsible parties. EPA can also take enforcement action against responsible parties to compel them to clean up sites. Other provisions of CERCLA require releases of hazardous substances over a specified amount ("reportable quantities") to be reported.

CERCLA is mostly an "after-the-fact" cleanup program, and there are no routine compliance monitoring inspections as in other programs. Sites are visited and environmental and other data are gathered for evaluation and assessment purposes, as well as to identify potential responsible parties. This information may ultimately be used in enforcement actions to recover the costs of cleanup or to compel cleanup by responsible parties.

Although CERCLA is not oriented to routine inspections of active industrial facilities, inspectors should be alert to signs of potential abandoned dump sites, spills, potential release of hazardous wastes, or other Superfund-type situations while they are out in the field, such as:

- Rusting drums and containers, evidence of spills, discolored vegetation, discolored water, foul-smelling lagoons

- Statements by facility personnel about how they handle wastes

- Records of spills or other releases of hazardous substances, or potential releases of hazardous substances

- Records of non-RCRA sites where hazardous substances have been stored, treated, or disposed

The investigator should determine, through records review, interviews, etc., whether all RCRA and CERCLA sites have been reported to the proper authorities. The investigator should also evaluate assessment and response programs at a facility, if this objective is within the scope of the audit.

Additionally, the facility should be evaluated concerning State and local requirements controlling past and current disposal of municipal waste, nonhazardous industrial waste, and construction debris. The information concerning such past disposal activities may lead to unreported RCRA and CERCLA sites.

The initial step in evaluating compliance with solid/hazardous waste requirements is to identify all present and past waste streams generated at the facility and determine which are regulated by Federal,* State,** or local regulations, licenses, and approvals. Preferably, this determination is initiated during background document review before the on-site facility audit and supplemented/modified using information obtained while on-site. All waste streams generated (even those that the generator claims are not regulated) must be evaluated for regulatory inclusion. This will allow the investigator to determine whether the generator has properly identified all regulated waste streams.

Once regulated waste is identified, the investigator can track the material from generation to final on-site disposition (on-site treatment/ disposal) or storage and transport for off-site disposal and determine compliance with applicable regulations. Throughout the investigation, the investigator must keep in mind that both past and present activities need to be evaluated for compliance with applicable regulations.

---

*   *Definitions, identification, and listing of Federally regulated waste are given in 40 CFR 260 and 261 and CERCLA § 101.*

** *Nonhazardous solid waste is usually regulated by the State and these regulations must be obtained to evaluate applicable facility activity.*

LABORATORY AND DATA QUALITY AUDITS

The purpose of laboratory evaluations and data quality assessment is to determine if all analytical and monitoring requirements have been met and to characterize data usability. The two approaches used are: (1) performance and (2) systems audits. This section discusses the approach to laboratory auditing under the CWA, RCRA waste handling, and RCRA groundwater regulations.

Performance audits are independent checks made to evaluate the quality of data produced by the total measurement system. This type of audit assesses the results and usually does not examine the intermediate steps to achieve these results. One example is the performance evaluation check sample which is used to validate calibration accuracy but usually not the overall effectiveness of the methodology. Another example is an audit of a particular measurement device using a reference device with known operational characteristics.

A systems audit typically involves an inspection of the components comprising the total measurement system. The Agency has certain expectations of the process used to sample, analyze, and report results. The systems audit is designed to objectively examine each important part of that process to determine deviations from required or recommended practice. The systems audit is more qualitative than the performance audit. A systems audit assesses such items as equipment, personnel, physical aspects, analytical and quality control procedures, quality assurance procedures, and other laboratory or measurement procedures. From a regulatory perspective, this type of audit may find noncompliance with equipment or procedural requirements, or even fraud.

Typically, a systems audit combined with performance audits will be conducted in order to extract the maximum amount of information.

A detailed list of items should be requested from the company and contract laboratory. This list should include:

- Standard Operating Procedures (SOPs)
- Quality Assurance Plan
- Personnel resumes
- Instrument maintenance and calibration records
- Monitoring data to be looked at

If performance evaluation samples are to be analyzed, these should be forwarded to the company at the earliest possible time. If preliminary data is available, it should be carefully examined for problems and if problems are found, a more careful examination of these areas can be made on-site.

During the on-site visit, every component of sample handling, sample analysis, and data reduction should be examined. The auditor starts with the laboratory supervisor and QA officer to verify that the information supplied on personnel training, quality assurance/quality control, and SOPs is correct. For each parameter determined, the individual or individuals who actually make that determination are interviewed. The analyst is asked to detail exactly what happens to each sample and demonstrate the use of equipment including instrument calibration. Checklists are prepared as an aid to the inspector. Bench data (initially recorded numbers, strip charts, etc.) is selected. Final results are calculated from the bench data by the inspector and compared with the results reported to the agency. On-site personnel will be asked to explain any discrepancies at this time. Other documents necessary to the case or as potential evidence are copied.

The final assessment and data quality determination is normally performed following the on-site audit. Critical data are re-examined for trends and anomolies. Where necessary, data is computerized and analyzed using statistical software packages. Techniques such as mass balance, solubility product determination, oxidation-reduction state consistency are used, where applicable, to indicate data problems. A propagation of error treatment may be used to establish data quality. Performance audit results are evaluated against reference database statistics. Tasks for common laboratory audits are:

NPDES (Water)

- Determine that the exact date, time, and person who takes each sample are recorded.

- Determine that the exact date, time, person, and method used for each type of determination are recorded.

- Inspect permit carefully to ensure that the permittee adheres to specified conditions.

- Ensure that methods used are in conformance with 40 CFR 136 unless alternate approval has been obtained.

- Ensure that proper chain-of-custody, accurate flow measurements, field preservation techniques, and instrument calibration procedures are practical.

RCRA Waste Handling

- Determine which parts of the regulations are applicable to the site.

- Determine which waste analysis plans (WAPs) were in effect during the time of records and evaluation.

- Determine that the WAPs meet the specifications of the regulation.

- Determine that each type of analysis specified in the WAPs is performed in accordance with the methodology specified and under the circumstances required.

- Determine that the methodology specified is adequate.

RCRA Groundwater

- Determine that the sampling and analysis plan (SAP) is adequate.

- Determine that the laboratory follows the methodology specified in the SAP.

- Determine that this methodology is adequate.

- Calculate detection limits to ensure that they are adequate for groundwater protection.

FIELD CITATIONS AND DOCUMENTATION

Field Citations

The use of field citations will be predicated on policies established by the individual program offices. They will be used if appropriate forms exist and the universe of violations for which they apply are well defined, unless specifically requested not to do so.

Other Facility Material

A serialized, document control system should be used to ensure that all facility documents are readily available when preparing the investigation report, and that all will be accounted for when the project is completed. All facility documents should be numbered, logged in for accounting purposes, provided to the appropriate investigation personnel, and ultimately filed or attached to the final investigation report, depending on the nature of the document. The team leader should have full responsibility for implementing this system for his/her particular investigation(s).

Documents received from a facility should be inventoried and a receipt for documents provided to the facility. EPA laws allow for the copying of documents. In some cases, facilities may not provide copies of requested documents so the investigators will have to provide their own document copying equipment (e.g., a rental portable copying machine). If the company provides copies, the investigator should offer and be prepared to pay a reasonable cost for each copy (see FOIA guidance/procedures for guidance on typical costs).

Project Logbook

The team leader should provide a bound logbook to every individual participating in the investigation. Each investigator should maintain his/her own investigation logbook, and they will form the basis for preparing the written investigation report. All logbooks issued by EPA are the property of EPA and should be turned over to the team leader for filing, after the final report is completed. In addition to documenting pertinent observations/findings/comments, logbooks should also include any *in situ* measurements and descriptive information relative to all sampling operations. Any change in the logbook should be initialed and dated. It is important to remember that logbooks can end up in court, and therefore, must only contain facts, figures, and observations.

Chain-of-Custody

The team leader should be responsible for taking an adequate number of sample tags and chain-of-custody forms for the on-site investigation. If possible, these should be in serialized order. He/she should distribute them to the team members on an as needed basis. Each investigator that collects samples must prepare and be custodian of all chain-of-custody records for those samples until they are returned to the laboratory. When chain-of-custody records are no longer needed by the sampler, the original copies should be given to the team leader who will then file them for possible future use.

Photographs

All photographs should remain in the possession of the investigator that took them. That person will also be responsible for properly labeling the photographs so that they can be attached to the investigation report, and making a copy of each so they can be included in the master project file. Each photograph should be given a separate number for identification purposes, when it is taken. Corresponding entries should also be made in the logbook. For each numbered photograph, the photographer should include the following information in his/her logbook:

- Name of photographer
- Date
- Time
- Subject of photograph
- Direction of photograph

Additional information on photographs/microfilm is presented in Appendix J.

INVESTIGATION REPORT

After the on-site investigation is completed, information obtained is further evaluated and findings/conclusions are developed. An inspection report is then written to present the findings, conclusions, and supporting information in a logical organized manner. Reports should be prepared and peer-reviewed before they are published in final form. The procedure involves developing a draft for internal review, then a subsequent, revised draft for external (client) review. Upon receipt of comments on the external review draft, a final report is prepared. The final report is the basis for follow-up activities or enforcement actions that might be initiated.

Inspection reports are prepared by the appropriate individual or project team member(s) under the direction of the team leader. All participants in the report preparation process must assure that their individual contributions to the report are accurate, relevant, objective, clear, fully supportable, and commensurate with Agency policy. Supporting information and documents used or referred to in the report are implicitly endorsed, unless disclaimed. Report authors are responsible for determining where such disclaimers are needed. Although the overall responsibility for the preparation and content of the reports rests with the project coordinator, team members are responsible for the quality, accuracy, and admissibility of information in the final report.

Many different formats are possible for the multi-media inspection report. Generally, reports for the Categories C and D inspections are longer and require more effort to produce a cohesive, readable document.

The potential audience for a multi-media inspection report may be diverse and includes not only technical peers but also managers, lawyers, judges, reporters, informed citizens, and other non-technical readers; the reports are written for this diverse audience.

Readability of the longer reports may be enhanced by organizing the report into two major sections: the Executive Summary and the Technical Report. The Executive Summary section clearly states inspection objectives, discusses relevant background information, summarizes inspection methods, and, as appropriate, presents conclusions regarding facility compliance which are supported by a brief summary of the findings. The Summary should include enough specifics to accurately determine whether a violation has occurred (e.g., "insufficient aisle space" is not all right; "aisle space less than 15 inches" is all right). The Technical Report section more comprehensively describes the inspection, giving specific details about the findings, including sample collection and analysis, and other pertinent aspects of the investigation. Findings in the Executive Summary must correlate to and be supported by discussion in the Technical Report.

Distribution of final reports is coordinated with the requesting office, program office, and Regional counsel. Reports containing confidential business information (CBI) may be subject to distribution restrictions. EPA reports containing material asserted to be CBI by the company may not be shared with non-Federal agencies without obtaining specific authorization from the company.[*]

OCCUPATIONAL SAFETY AND HEALTH ADMINISTRATION

The Secretary of Labor and Administrator of EPA signed a Memorandum of Understanding (MOU) on November 23, 1990 with the goal of establishing a program for improved environmental and workplace health and safety. Implementation of the program is to be coordinated primarily by the Occupational Safety and Health Administration (OSHA)

---

[*]   *Restrictions on distribution of CBI information are presented in 40 CFR Part 2.*

and EPA Office of Enforcement. Although the two agencies have worked cooperatively together in the past on a number of issues and investigations, no comprehensive structure existed to focus that cooperative effort nationally. Having such a comprehensive structure is particularly critical, given the need to assure the most effective use of limited Federal resources and potential overlapping EPA-OSHA responsibilities.

The MOU provides for coordinated and joint inspections of facilities believed to be in violation of Federal workplace or environmental standards, facilitates the exchange of technical information, computer data bases, and other information to allow for better targeting of inspections, and provides for cross-training programs.

The MOU requires that a number of specific actions be taken, including the development of a workplan for 1991 with subsequent annual workplans to be developed by the beginning of each succeeding fiscal year. Separate agreements and data exchange will also be developed in the future.

Furthermore, all Agency investigators should be aware of OSHA requirements and be alert for potential violations of OSHA requirements. Team leaders should be aware of appropriate procedures to refer potential violations to OSHA.

REFERENCES

REFERENCES

1.   U.S. Department of Health and Human Services. NIOSH/OSHA/USCG/EPA. 1985. <u>Occupational Safety and Health Guidance Manual for Hazardous Waste Site Activities</u>. U.S. Government Printing Office.

2.   U.S. Department of Health and Human Services. 1987. <u>NIOSH Respirator Decision Logic</u>. DHHS Publication No. 87-108.

3.   U.S. Department of Justice. James C. Cissel, United States Attorney, Southern District of Ohio. 1980. <u>Proving Federal Crimes</u>. U.S. Government Printing Office.

4.   U.S. Environmental Protection Agency. A.D. Schwope, P.P. Costas, J.O. Jackson, J.O. Stull, D.J. Weitzman. 1987. <u>Guidelines for the Selection of Chemical Protective Clothing</u>. Prepared for the Office of Administration.

5.   U.S. Environmental Protection Agency. Enforcement Specialist Office. 1987. <u>Federal Environmental Criminal Laws and Ancillary U.S. Code Violations</u>.

6.   U.S. Environmental Protection Agency. Hazardous Waste Ground-Water Task Force. 1986. <u>Protocol for Ground-Water Evaluations</u>. OSWER 9080.0-1.

7.   U.S. Environmental Protection Agency. National Enforcement Investigations Center. 1981. <u>NEIC Manual for Groundwater/Subsurface Investigations at Hazardous Waste Sites</u>. EPA-330/9-81-002.

8.   U.S. Environmental Protection Agency. National Enforcement Investigations Center. 1978 [Revised August 1991]. <u>NEIC Policies and Procedures Manual</u>.

9.   U.S. Environmental Protection Agency. Occupational Health and Safety Staff/Office of Administration. 1986. <u>Occupational Health and Safety Manual</u>. Transmittal 1440.

10. U.S. Environmental Protection Agency. Office of Enforcement and Compliance Monitoring. 1989. <u>Basic Inspectors Training Course: Fundamentals of Environmental Compliance Inspections</u>.

11. U.S. Environmental Protection Agency. Office of Federal Activities. 1989. <u>Generic Protocol for Environmental Audits of Federal Facilities</u>. EPA/130/4-89/002.

REFERENCES (cont.)

12. U.S. Environmental Protection Agency. Office of Ground-Water Protection. 1987. <u>Guidelines for Delineation of Wellhead Protection Areas</u>.

13. U.S. Environmental Protection Agency. Office of Pesticides and Toxic Substances. 1989. <u>Pesticides Inspection Manual</u>.

14. U.S. Environmental Protection Agency. Office of Solid Waste. 1984. <u>Waste Analysis Plans: A Guidance Manual</u>. EPA/530-sw-84-012.

15. U.S. Environmental Protection Agency. Office of Toxic Substances. 1984. <u>TSCA Confidential Business Information Security Manual</u>.

16. U.S. Environmental Protection Agency. Office of Waste Programs Enforcement. 1991. <u>Conducting RCRA Inspections at Mixed Waste Facilities</u>. OSWER 9938.9.

17. U.S. Environmental Protection Agency. Office of Waste Programs Enforcement. 1989. <u>Hazardous Waste Incinerator Inspection Manual</u>. OSWER 9938.6.

18. U.S. Environmental Protection Agency. Office of Waste Programs Enforcement. 1988. <u>Hazardous Waste Tank Systems Inspection Manual</u>. OSWER 9938.4.

19. U.S. Environmental Protection Agency. Office of Waste Programs Enforcement. 1988. <u>RCRA Ground-water Monitoring Systems: RCRA Comprehensive Ground-water Monitoring Evaluation Document, Operation and Maintenance Inspection Guide, RCRA Laboratory Audit Inspection Guidance</u>. OSWER 9950.4

20. U.S. Environmental Protection Agency. Office of Waste Programs Enforcement. 1988. <u>RCRA Inspection Manual</u>. OSWER 9938.2a.

21. U.S. Environmental Protection Agency. Office of Waste Programs Enforcement. <u>RCRA Technical Case Development Guidance Document</u>. OSWER 9938.3.

22. U.S. Environmental Protection Agency. Pesticides and Toxic Substances Enforcement Division. 1990. <u>Toxic Substances Control Act Inspection Manual</u>.

23. U.S. Environmental Protection Agency. Region X. 1990. <u>KISS Manual</u>. Prepared for the Basic Inspector Training Course.

REFERENCES (cont.)

24.   U.S. Environmental Protection Agency. Solid Waste and Emergency Response. 1989. <u>Land Disposal Restrictions: Inspection Manual</u>. OSWER 9938.1a.

25.   U.S. Environmental Protection Agency. Solid Waste and Emergency Response. 1991. <u>Land Disposal Restrictions: Summary of Requirements</u>. OSWER 9934.0-1a.

26.   U.S. Environmental Protection Agency. Office of Water Enforcement and Permits. May 1988. <u>NPDES Compliance Inspection Manual</u>.

APPENDICES

A   SUMMARY OF POLLUTION CONTROL LEGISLATION
B   MULTI-MEDIA INSPECTIONS - DEFINITIONS AND TRAINING
C   TEAM LEADER RESPONSIBILITIES AND AUTHORITIES
D   TYPES OF INFORMATION
E   SOURCES OF INFORMATION
F   PROJECT PLAN AND SAFETY PLAN EXAMPLES
G   MULTI-MEDIA INVESTIGATION EQUIPMENT CHECKLIST
H   SUGGESTED RECORDS/DOCUMENT REQUEST
I   MULTI-MEDIA CHECKLISTS
J   INVESTIGATION AUTHORITY/SUMMARY OF MAJOR
          ENVIRONMENTAL ACTS
K   EVIDENTIARY PROCEDURES FOR PHOTOGRAPHS/MICROFILM
L   KEY PRINCIPLES AND TECHNIQUES FOR INTERVIEWING
M   SAMPLING GUIDELINES
N   LAND DISPOSAL RESTRICTIONS PROGRAM
O   SOURCES SUBPART (40 CFR PART 60) EFFECTIVE DATE OF
          STANDARD AIR POLLUTANTS SUBJECT TO NSPS
P   TSCA FORMS
Q   TSCA SECTIONS 5 AND 8 GLOSSARY OF TERMS AND ACRONYMS

APPENDIX A

SUMMARY OF POLLUTION CONTROL LEGISLATION

Appendix A

SUMMARY OF POLLUTION CONTROL LEGISLATION

This appendix is a synopsis of the Federal approach to environmental regulation, EPA enforcement remedies and a summary of each of the major pollution control acts: the Clean Air Act (CAA), the Clean Water Act (CWA), the Resource Conservation and Recovery Act (RCRA), the Comprehensive Environmental Response, Compensation and Liability Act (CERCLA/Superfund), the Toxic Substances Control Act (TSCA), the Federal Insecticide, Fungicide and Rodenticide Act (FIFRA), the Safe Drinking Water Act (SDWA), and the Emergency Planning and Community Right-to-Know Act (EPCRA). Because these laws and the regulations promulgated thereunder typically are very complex and are continually being modified, the investigator should carefully review the specific provisions which apply to the operations of the facility before conducting an inspection.

GENERAL FEDERAL APPROACH TO ENVIRONMENTAL REGULATION

National standards are established to control the handling, emission, discharge, and disposal of harmful substances. Waste sources must comply with these national standards whether the programs are implemented directly by EPA or delegated to the States. In many cases, the national standards are applied to sources through permit programs which control the release of pollutants into the environment. EPA establishes the Federal standards and requirements and approves State programs for permit issuance.

The States can set stricter standards than those required by Federal law. Some of the larger programs which have been delegated by EPA to qualifying States are the National Emissions Standards for Hazardous Air Pollutants (NESHAP), and the Prevention of Significant Deterioration (PSD) permits under the CAA, the Water Quality Standards, and the National Pollution Discharge Elimination System (NPDES) programs under the

CWA, the Hazardous Waste Program under RCRA, and the Drinking Water and Underground Injection Control (UIC) programs under the SDWA. Conversely, TSCA is administered entirely by the Federal government; although, States may have their own program regulating PCBs and asbestos.

EPA ENFORCEMENT OPTIONS

- Issuance of an Administrative Compliance Order, sometimes preceded by a Notice of Violation[*] - A Compliance Order will specify the nature of the violation and give a reasonable time for compliance. The order, if violated, can lead to enforcement action pursuant to the civil and/or criminal process of environmental laws.

- Issuance of an administrative complaint for civil penalties - Parties named in such complaints must be given notice and an opportunity for a hearing on the alleged violations before a penalty can be assessed by EPA.

- Under certain statutes (e.g., SDWA) EPA may take whatever action is necessary to protect the public health, in emergency situations, without first obtaining a judicial order.

- EPA generally may go directly to Federal court seeking injunctive relief or a civil penalty without using administrative procedures. EPA also may obtain an emergency restraining order halting activity alleged to cause "an imminent and substantial endangerment" or "imminent hazard" to the health of persons.

- EPA may go directly to Federal court seeking criminal sanctions without using administrative procedures. Criminal penalties are available for "knowing" or for "willful" violations.

---

[*] *A concise written statement with factual basis for alleging a violation and a specific reference to each regulation, act, provision or permit term allegedly violated*

In addition, EPA may suspend and/or debar a company or party that fails to comply with the environmental statutes by preventing it from entering into Federal contracts, loans, and grants. In cases where the party has been convicted of certain criminal offenses under the CAA or CWA, Federal agencies are expressly prohibited from entering into contracts, etc., with that entity.

CLEAN AIR ACT

The Clean Air Act (CAA), as amended in 1990, is one of the most comprehensive and ambitious environmental statutes ever enacted. Through it's various programs, it is intended to protect human health and the environment by reducing emissions of specified pollutants at their sources, thus allowing the achievement and maintenance of maximum acceptable pollution levels in ambient air. The CAA also contains provisions which seek to prevent presently existing unpolluted areas from becoming significantly polluted in the future. Regulations implementing the multitude of amendments enacted in 1990 will be promulgated pursuant to statutory deadlines for many years to come. Where regulations under the amendments have not yet been promulgated, requirements which existed prior to the 1990 amendments will continue to be enforceable until amended or new requirements are promulgated.

National Ambient Air Quality Standards (NAAQS)

As in prior versions of the CAA, Section 109 continues to require that EPA establish NAAQS to protect public health and welfare from air pollutants. These standards will apply in all areas of the country. "Primary" NAAQS must be designed to protect human health while building in an adequate margin for safety, whereas "secondary" NAAQS protect public welfare, including wildlife, vegetation, soils, water, property, and personal comfort. EPA has promulgated NAAQS for six air pollutants (criteria pollutants): ozone, carbon monoxide (CO), particulate matter (PM-10), sulfur dioxide (SO_2), nitrogen dioxide and lead.

State Implementation Plans (SIPs)

The States, through adoption of plans known as SIPs, are required to establish procedures to achieve and maintain all NAAQS promulgated by EPA. EPA has designated 247 Air Quality Control Regions (AQCRs). Each AQCR has been evaluated to determine whether the NAAQS for each of the criteria pollutants has been met. AQCRs which do not meet the NAAQS for any of the criteria pollutants are designated as "non-attainment" for those pollutants. Thus, one AQCR may be attainment for some pollutants and non-attainment for others.

The SIP program dates back to the 1970 Clean Air Act, which required States to promulgate SIPs by 1972 to assure attainment of air quality standards by 1977. Having not met that goal, the 1977 amendments continued the program, requiring additional controls designed to achieve attainment by 1982, or at the latest 1987. While the goals of the SIP program were again not reached, the 1990 amendments have further continued the effort, adding several requirements which may increase the effectiveness of the program, including the use of modeling and other specified analytical techniques to demonstrate the ability to achieve attainment and a wide range of specified control requirements. An additional advantage of the new SIP program is that under the 1990 amendments it is no longer the primary mechanism for implementation of NAAQS. Instead, the comprehensive permit program under the 1990 amendments will assume a large part of that burden by detailing the specific requirements applicable to individual sources, thereby resolving any uncertainties as to what requirements are applicable.

A SIP must contain strategies designed to meet targets for attainment of any NAAQS for which an area is non-attainment by prescribed dates. SIPs must meet Federal requirements, but each State may choose its own mix of emissions for stationary and mobile sources to meet the NAAQS. The deadline for attainment of primary NAAQS is no more than five (5) years after the area was designated non-attainment, although EPA has the authority to extend the deadline for up to five (5) additional years. Attainment for secondary NAAQS must be achieved "as expeditiously as practicable."

In order to accomplish attainment, States must impose controls on existing sources to reduce emissions to the extent necessary to ensure achievement of the NAAQS. In attainment areas, new sources and those which are undertaking modifications which will increase emissions by more than a *de minimis* amount must obtain State construction permits after demonstrating that anticipated emissions will not exceed allowable limits. In non-attainment areas, emissions from new or modified sources must be offset by emissions reductions from existing sources.

Each State must submit a proposed SIP to EPA for approval within three (3) years of designation as non-attainment. Failure to submit a SIP, failure to submit an adequate SIP or failure to implement a SIP may subject a State to the imposition of sanctions such as increased offset ratios for stationary sources, prohibition of Federal highway grants or a ban on air quality grants. For ozone non-attainment areas, failure to attain the NAAQS will result in reclassification of the area, thus imposing more stringent control requirements and imposition of financial penalties on stationary sources in severe or extreme non-attainment areas. Where an acceptable SIP is not submitted by a State, EPA will be required to propose and enforce a Federal Implementation Plan for that State. EPA and the States have concurrent enforcement authority for SIPs.

Deadlines and control requirements imposed upon non-attainment areas vary depending upon the severity of the existing air pollution problem, with correspondingly more stringent control requirements and longer deadlines applying to more polluted areas. The CAA creates five (5) classes of ozone non-attainment and two (2) categories each for carbon monoxide and PM-10 non-attainment areas.

Prevention of Significant Deterioration (PSD)

The purpose of PSD, which remains largely unchanged by the 1990 amendments, is to avoid significant future degradation of the nation's clean air areas. A clean air area is one where the air quality is better than the ambient primary or secondary standard. Designation is pollutant specific so that an area can be non-attainment for one pollutant but clean for another. PSD applies only to new and modified sources in attainment areas. Clean air areas are divided into three categories: Class I includes wilderness areas and other pristine areas, where only minor air quality degradation is allowed; Class II includes all other attainment and non-classified areas where moderate degradation is permitted; and Class III includes selected areas that States designate for development where substantial degradation is permitted. In no case would PSD allow air quality to deteriorate below secondary NAAQS.

"Baseline" is the existing air quality for the area at the time the first PSD permit is applied for. "Increments" are the maximum amount of deterioration that can occur in a clean air area over baseline. Increments in Class I areas are smaller than those for Class II areas and Class II increments are smaller than those for Class III areas. For purposes of PSD, a major emitting source is one which falls within 28 designated categories and emits or has the potential to emit more than 100 tons per year of the designated air pollutant. A source that is not within the 28 designated categories is a major source if it emits more than 250 tons per year. Modifications to major sources that will result in a "significant net emissions increase" of any regulated pollutant are also subject to PSD. The amount of emissions which qualifies as significant varies for the regulated pollutants.

Under this program, new "major stationary sources" and "major modifications" to such sources located in attainment areas must obtain a permit before beginning construction. Permit requirements include installation of Best Available Control Technology (BACT) for each regulated pollutant emitted in significant amounts, assurance that the new emissions will not exceed NAAQS or any maximum allowable "increment" for the area, and assurance that the new emissions will not adversely impact any other air quality related values, such as visibility, vegetation or soils.

Hazardous Air Pollutants

Prior to the enactment of the 1990 amendments, Section 112 of the CAA required the establishment of National Emission Standards for Hazardous Air Pollutants (NESHAPs) to regulate exposure to dangerous air pollutants that are so localized that the establishment of NAAQS is not justified. NESHAP standards were to be based on health effects, with strong reliance on technological capabilities. They applied to both existing and new stationary sources. During the 20 years in which this program existed, effective regulations for only seven (7) substances were enacted: benzene, beryllium, asbestos, mercury, vinyl chloride, arsenic, and radionuclide emissions.

As rewritten in 1990, the goal of Section 112 remains the same - to protect public health and the environment from toxic air pollutants for which NAAQS will not be established. While the new program requires standards to be set for categories and subcategories of sources that emit hazardous air pollutants, rather than for the air pollutants themselves as under the NESHAP program, the seven (7) NESHAPs promulgated prior to the amendments will generally remain applicable until they are revised pursuant to the timetables established in the new Section 112.

Under the 1990 amendments, two types of sources have been identified for purposes of establishing emission standards - "major sources", which include stationary sources or a group of stationary sources within a contiguous area and under common control that emit or have the potential to emit 10 tons per year of a single listed hazardous air pollutant or 25 tons per year of any combination of listed hazardous air pollutants; and "area sources", which include numerous small sources that may cumulatively produce significant quantities of a pollutant resulting in a threat of adverse effects on human health or the environment. An initial list of 189 air pollutants requiring regulation was established by Congress and EPA has been tasked with the responsibility for establishing lists of categories and subcategories of major sources and area sources subject to emission standards.

Major Sources

EPA is required to set technology-based standards for sources of the listed pollutants which are designed to achieve "the maximum degree of reduction in emissions" (Maximum Achievable Control Technology - MACT) while taking into account costs and other health and environmental impacts. The standards for new sources "shall not be less stringent than the most stringent emissions level that is achieved in practice by the best controlled similar source" in the same category or subcategory. For existing sources, the standards may be less stringent than those for new sources, but in most circumstances must be no less stringent than the emissions control achieved by the best performing 12% of sources in the category or subcategory [or five (5) sources in a category with less than 30 sources]. Existing sources are given three (3) years following the promulgation of standards to achieve compliance, with the possibility of a one- (1)-year extension. Sources that voluntarily reduce emissions by 90% before an applicable MACT is proposed (95% for hazardous particulates) may be granted one (1) six- (6)-year extension from the MACT. Solid waste incinerators will be required to comply with both these hazardous air pollutant standards and the new source performance standards to be promulgated pursuant to Section 111 of the CAA.

The second provision of Section 112 relating to major sources sets health-based standards to address situations in which a significant residual risk of adverse health effects or a threat of adverse environmental effects remains after installation of MACT. Within six (6) years of enactment of the CAA, and after consultation with the Surgeon General and opportunity for public comment, EPA must report to Congress regarding the public health significance of the residual risks, technologically and commercially available methods and costs of reducing such risks and legislative recommendations to address such residual risks. If Congress does not act on the recommendations submitted by EPA, the EPA must issue residual risk standards for listed categories and subcategories of sources as necessary to protect public health with an ample margin of safety or to prevent adverse environmental effects.

Area Sources

The goal of the area source program is to reduce the incidence of cancer attributable to stationary area sources by at least 75% through a comprehensive national strategy for emissions control in urban areas. By November 15, 1995, EPA is required to identify the 30 hazardous air pollutants emitted from area sources that pose the most significant risks to public health in the largest number of urban areas and the source categories and subcategories of those pollutants. Area sources representing at least 90% of the emissions of the 30 identified pollutants will be subject to regulations to be promulgated by EPA by November 15, 2000.

Prevention of Sudden Catastrophic Releases

As added by the 1990 amendments, Section 112 of the CAA imposes a general duty on owners and operators of stationary sources which handle hazardous substances to: identify hazards which may result from releases, design and maintain safe facilities, take action to prevent releases, and minimize the consequences of accidental releases that do occur. It also requires EPA to promulgate a list of substances which, in the event of an accidental release, may reasonably be anticipated to cause death, injury, or serious adverse health and environmental effects, as well as threshold quantities for each of those regulated substances. Requirements for release prevention, detection and correction of regulated substances must be promulgated by EPA. Among the requirements will be preparation and implementation of risk management plans by owners and operators of facilities with regulated amounts greater than the threshold quantity.

Emergency policies and the opportunity to secure relief in the district courts is provided to EPA to protect against an imminent and substantial endangerment to health or the environment as a result of an actual or threatened release of a regulated substance. An independent, five-member Chemical Safety and Hazard Investigation Board to be appointed by the President will investigate any accidental release that results in a fatality, serious injury or substantial property damage, and will issue a report to EPA and OSHA recommending regulations for preparing risk management plans and general requirements for preventing and mitigating the potential adverse effects of accidental releases.

New Source Performance Standards (NSPS)

With the exception of the extension and establishment of deadlines for EPA's proposal of various regulations, the NSPS program remains largely unchanged by the 1990 amendments. NSPS establishes nationally uniform, technology-based standards for categories of new industrial facilities by providing maximum emission levels for new or extensively modified major stationary sources. The emission levels are determined by the best "adequately demonstrated" continuous control technology available, taking costs into account. Regulations for source categories listed prior to November 15, 1990, must be proposed in phases, beginning on November 15, 1992. Standards for new categories listed after November 15, 1990, must be proposed within one (1) year of listing and must be finalized within one (1) year after proposal.

The owner or operator of a new or extensively modified major source is required to demonstrate compliance with an applicable NSPS within 180 days of initial start-up of the facility and at other times required by EPA. Primary authority for enforcement of NSPS lies with EPA unless that authority is delegated to States, in which case EPA and the States have concurrent enforcement authority.

Emission Standards for Mobile Sources

Section 202 of the CAA directs EPA to regulate air pollutants emitted by motor vehicles which "cause, or contribute to air pollution which may reasonably be anticipated to endanger public health or welfare." In response, the Agency has set standards governing motor vehicle emissions of carbon monoxide, hydrocarbons, oxides of nitrogen and particulates. These standards have given rise to the emission control systems that first appeared in automobiles in the early 1970s. The CAA generally prohibits the removal (or rendering inoperative) of any emission control device that was installed by the vehicle manufacturer in order to meet the applicable emission standards. Most States have enacted similar laws enforcing this prohibition and/or have incorporated such prohibitions as part of their SIP.

The CAA provides EPA with the authority to control or prohibit the use of fuels which pose a public health risk or which "impair to a significant degree the performance of any emission control device or system." The Agency's regulations are based upon both of these rationales. (The best example of this is the regulations governing the lead content of gasoline.) Enforcement of the fuel standards is achieved through a combination of Federal and State efforts, and is based, in part, upon SIP provisions and/or State laws.

The 1990 amendments tightened emission standards for both heavy duty and light duty vehicles. These standards take effect at different times for different types of vehicles, beginning in 1994.

Beginning in 1995, "reformulated" gasoline limiting emissions of air pollutants must be sold in the nine worst ozone non-attainment areas (Los Angeles, San Diego, Baltimore, Philadelphia, New York, Hartford, Chicago, and Milwaukee). Other ozone non-attainment areas may elect to use reformulated gasoline as it becomes more widely available.

Two "alternative fuel" programs are expected to reduce emissions in the most seriously polluted areas. California will develop a program requiring introduction of low emission vehicles and ultra low emission vehicles beginning in 1996. Additionally, in more than 20 metropolitan areas, fleets of 10 or more vehicles are required to phase in usage of "clean fuel vehicles" beginning in 1998.

<u>Acid Rain Control</u>

The acid rain program, added to the CAA by the 1990 amendments, primarily impacts emissions of sulfur dioxide (SO_2) and nitrogen oxides (NOx) from powerplants. It requires establishment of specific NOx emission rate limitations and implementation of a two-phase reduction of SO_2 emissions by the year 2000.

As determined by a formula based on 1985 SO_2 emissions and 1985 to 87 annual average fuel consumption for individual powerplants, marketable allowances are to be allocated to powerplants by EPA. As a general rule, only powerplants operating prior to November 15, 1990, will receive an allocation of allowances for SO_2 emissions. New units which begin operation after November 15, 1990, will be required to obtain offset allowances from existing facilities in order to continue operation after the year 2000. Allowances can be bought, sold, shared with other regulated units or banked for future use. Facilities are not allowed to emit more SO_2 than the amount for which they hold allowances.

Penalties will be imposed at the rate of $2,000 per excess ton of SO_2 on any facility which does not have sufficient allowances to cover its SO_2 emissions. Additionally, any offending facility will be required to reduce its SO_2 emissions by one ton the following year for each ton of excess SO_2 emitted.

Stratospheric Ozone Protection Program

Another program added by the 1990 amendments is designed to protect the earth's stratospheric ozone layer by phasing out production and use of ozone-depleting substances, providing limited exemptions for uses such as medical, aviation, and fire-suppression. Production of Class I substances, those with the greatest depleting potential, will generally be prohibited after January 1, 2000. Production of Class II substances, those with less depleting potential, will be prohibited after January 1, 2030. EPA is required to promulgate rules governing issuance of production allowances for Class I and II substances and banning the production, after November 15, 1992, of non-essential products that release Class I substances.

Permitting

Under the Clean Air Act as it existed prior to the 1990 amendments, permits were required for only a limited number of facilities. While these requirements will continue to apply until new regulations are promulgated, the 1990 amendments have expanded the permit program to require most regulated stationary sources to have permits.

The new permitting program, patterned after the NPDES program, is designed to consolidate all operation and control requirements in one permit. However, unlike the NPDES program which focuses on individual sources within a facility, air permits are expected to be issued to a facility as a whole. As a result of this comprehensive program, greater consistency is expected and facilities will not be subjected to conflicting requirements.

Permits are required for any facility that qualifies as a "major source", which generally includes any source emitting more than 100 tons of pollutants per year, but extends to smaller sources in the more seriously polluted non-attainment areas. Permits are also required for major sources and area sources subject to regulation for emissions of hazardous air pollutants under Section 112 and all sources subject to NSPS.

Regulations establishing the numerous requirements for state permit programs must be promulgated by EPA by November 15, 1991. States must then develop and submit to EPA an operating permit program for approval by November 15, 1993. If all or any part of the program is disapproved by EPA, the States must correct the deficiencies and resubmit the program. A failure to timely submit a program or correct deficiencies will result in sanctions. If a program is not completely approved within 2 years after initial submission of the program to EPA or by November 15, 1995, whichever is earlier, EPA must promulgate and administer a permit program for the State.

Permit applications must be filed within 12 months after the permit program takes effect and must include a compliance plan for the facility. The permits, as issued, will contain enforceable emission limitations and standards, a schedule of compliance, and compliance certification, inspection, entry, monitoring and reporting requirements. Compliance with a permit will, to some extent, shield a source from enforcement actions. The extent of the protection provided by compliance will be governed by EPA's permitting regulations to be promulgated by November 15, 1991.

<u>Enforcement</u>

The 1990 Amendments greatly expanded enforcement options available under the CAA and impose heavy penalties, both civil and criminal, for violations of the Act.

Administrative penalties of up to $25,000 per day, to a maximum of $200,000, may be imposed by EPA for violations of any requirement, prohibition, permit, rule, or order without the initiation of a court proceeding. These penalties can be overturned only if, on judicial review, they are not supported by substantial evidence. Field investigators are also authorized to issue "field citations" imposing penalties of up to $5,000 per day per violation for minor violations observed while on site. Administrative orders requiring specific actions to comply with the CAA may be issued where compliance can be achieved within 1 year. Additionally, private citizens are now authorized to bring citizen suits seeking civil penalties for violations of the CAA where neither EPA nor the State is "diligently prosecuting a civil action" to require compliance, or seeking to compel EPA to discharge a non-discretionary duty, such as promulgating regulations by statutory deadlines.

Knowing violations of many provisions of the CAA qualify as felony crimes, punishable by fines for individuals of up to $250,000 and imprisonment up to 5 years, with each day counting as a separate violation. Fines for corporations may be up to $500,000 per day per violation. Penalties may be doubled for second convictions. The negligent release of a hazardous air pollutant or extremely hazardous substance under SARA that puts another person in imminent danger of death or serious bodily injury is punishable by fines and imprisonment for up to 1 year, while knowingly releasing such substances is punishable by up to $250,000 per day and 15 years imprisonment for individuals and up to $1 million per day for businesses.

EPA is also authorized to pay a "bounty" of up to $10,000 for information leading to a criminal conviction or a judicial or administrative civil penalty for violations of the CAA.

The CAA contains a presumption that once a violation which is likely to be of a continuing nature is proven, the violation is presumed to continue until full compliance is achieved unless the defendant can prove that the violation ceased.

CLEAN WATER ACT (FEDERAL WATER POLLUTION CONTROL ACT)

Through the 1950s and 1960s, emphasis was on the States setting ambient water quality standards and developing plans to achieve these standards. In 1972, the Federal Water Pollution Control Act was significantly amended. These changes emphasized a new approach, combining water quality standards and effluent limitations (i.e., technology-based standards). The amendments called for compliance by all point-source discharges with the technology-based standards. A strong Federal enforcement program was created and substantial monies were made available for construction of sewage treatment plants. The Federal Water Pollution Control Act was amended in 1977 to address toxic water pollutants and in 1987 to refine and strengthen priorities under the Act as well as enhance EPA's enforcement authority. Since the 1977 amendments, the Federal Water Pollution Control Act has been commonly referred to as the Clean Water Act (CWA).

State Water Quality Standards and Water Quality Management Plans

Section 303 of the CWA authorizes the States to establish ambient water quality standards and water quality management plans. If national technology standards are not sufficient to attain desired stream water quality, the State shall set maximum daily allowable pollutant loads (including toxic pollutants) for these waters and, accordingly, determine effluent limits and compliance schedules for point sources to meet the maximum daily allowable loads.

The National Pollutant Discharge Elimination (NPDES) Program

This program was established by Section 402 of the CWA and, under it, EPA and approved States have issued more than 50,000 NPDES permits. Permits are required for all point sources from which pollutants are discharged to navigable waters. An NPDES permit is required for any __direct__ discharge from new or existing sources. __Indirect__ discharges through POTWs are regulated under a separate program (see discussion of pretreatment standards below). In 1979 and 1980, the permit program was revised and one of the new features was the use of __Best Management Practices (BMPs)__ on a case-by-case basis to minimize the introduction of toxic and hazardous substances into surface waters. BMPs are industry practices used to reduce secondary pollution (e.g., raw material storage piles shall be covered and protected against rain and runoff). BMPs do not have numerical limits and, therefore, are different from effluent limits.

Section 304 of the CWA sets restrictions on the amount of pollutants discharged at industrial plant outfalls. Amounts are usually expressed as weight per unit of product (i.e., 0.5 lb/1,000 lb product manufactured). The standards are different for each industry. Effluent guidelines are applied to individual plants through the NPDES permit program.

There are three levels of technology for existing industrial sources: Best Practicable Control Technology (BPT), Best Conventional Technology (BCT), and Best Available Technology Economically Achievable (BAT). Under the 1972 Act, BPT was intended to be put in place by industry in 1977 and BAT in 1983. These timetables have been modified by subsequent amendments.

The 1987 CWA Amendments modified the compliance deadlines for the following:

- BPT limits requiring a substantially greater level of control based on a fundamentally different control technology

- BAT for priority toxic pollutants

- BAT for other toxic pollutants

- BAT for nonconventional pollutants

- BCT for conventional pollutants

For each technology the new deadline requires compliance "as expeditiously as practicable, but in no case later than 3 years after the date such limitations are promulgated. . .and in no case later than March 31, 1989."

New Source Performance Standards (NSPS) are closely related to BAT for existing sources but are not quite the same. NSPS are different for each industrial category. These standards must be achieved when the new industrial source begins to discharge. NSPS permits will be effective for a period of 10 years vs. 5 years or less for the BPT and BAT-type permits. This 10-year protection insulates against change in BCT or BAT requirements but does not hold against Section 307(a) toxic pollutant standards or against "surrogate" pollutants that are used to control hazardous or toxic pollutants.

A permit application must be made. Adequate information must be submitted including basic facility descriptions, SIC codes, regulated activities, lists of current environmental permits, descriptions of all outfalls, drawings, flows, treatment, production, compliance schedules, effluent characteristics, use of toxics, potential discharges, and bio-assay toxicity tests performed.

Applicants must conduct analytical testing for pollutants for BOD, COD, TOC, TSS, ammonia, temperature, and pH. The applicant, if included within any of the 34 "primary industry" categories, must sample for all toxic metals, cyanide, and phenols given in EPA Application Form 2C and for specified organic toxic pollutant fractions.

The applicant must list hazardous substances believed to be present at the industrial plant. Testing is not required but analytical results must be provided, if available.

NPDES Permit

The NPDES permit, issued by EPA or the State, enforces Federal effluent limitations promulgated for individual industrial categories; NSPS; toxic effluent standards; State water quality standards under Section 303 of the CWA, if any are applicable; and hazardous substances otherwise regulated under Section 311 of the CWA that may be incorporated under the NPDES permit instead. Permit elements include the amount of pollutants to be discharged expressed in terms of average monthly and maximum daily loads; compliance schedules, _if_ applicable standards cannot be met now; and monitoring, testing, and reporting requirements.

Routine Noncompliance Reports - The Discharge Monitoring Form

The Discharge Monitoring Report (DMR) gives a summary of the discharger's records on a monthly or quarterly basis for flow measurement, sample collection, and laboratory analyses. Noncompliance reports must be submitted quarterly on the cause of noncomplying discharges, period of noncompliance, expected return to compliance and plans to minimize or eliminate recurrence of incident.

Emergency Reporting

- _Health_: EPA shall be notified within 24 hours of noncompliance involving discharge of toxic pollutants, threat to drinking water, or injury to human health.

- _Bypass_: Noncompliance due to intentional diversion of waste shall be reported promptly to the permitting agency and may be permissible if essential to prevent loss of life or serious property damage.

- <u>Upset</u>: Temporary noncompliance due to factors beyond the reasonable control of the permittee shall be promptly reported to the agency.

The 1987 CWA Amendments establish a schedule for the regulation of municipal and industrial stormwater discharges under NPDES permits. Initially, (before October 1, 1992), only major dischargers and those who are significant contributors of pollutants will be required to obtain permits.

<u>Pretreatment Standards for Indirect Discharges to Publicly-Owned Treatment Works</u>

<u>Coverage</u>

New and existing industrial users who discharge to POTWs are subject to general and categorical pretreatment standards. The categorical standards are primarily directed to control of toxic pollutants in specific industries. Note that localities with approved pretreatment programs may have imposed local limits, which are enforceable by EPA.

<u>Requirements</u>

- <u>General Pretreatment Standards</u>

  Prohibit fire or explosion hazards, corrosivity, solid or viscous obstructions, "slug" discharges, and heat sufficient to inhibit biological activity at POTWs.

- <u>Categorical Standards</u>

  - Standards to be expressed as concentration limits or mass weight per unit of production.

  - Source must be in compliance 3 years after promulgation of standards.

- Variances can be obtained for fundamentally different factors or if industrial pollutants are consistently being removed by POTW.

- **Reports**

  Users must provide appropriate agency (EPA, State, or POTWs having approved pretreatment programs) with basic information; SIC code; average and maximum daily discharge; characteristics or pollutants, applicable standards and certification whether standards are being met and, if not, what pretreatment is necessary; and a compliance schedule.

- **Monitoring, Sampling, and Analysis**

  Users shall submit sampling data for each regulated pollutant in discharge.

- **Progress Reports**

  Reports and information shall be submitted at 6-month intervals.

Nonpoint Source Pollution Control

Section 208 of the CWA provides for control of nonpoint source pollution and directs States to establish planning bodies to formulate area-wide pollution control plans. NPDES permits cannot be issued where the permit may conflict with an approved Section 208 plan.

The 1987 CWA Amendments require States or EPA to develop nonpoint source management programs under Section 319.

Municipal and Industrial Stormwater Discharges

For some time there has been considerable debate over whether permits should be required for stormwater discharges from point sources, particularly those municipal or industrial discharges which may well contain toxic and other pollutants. The 1987 Amendments provide that five (5) types of stormwater discharges will be regulated under NPDES:

1. Discharges which have NPDES permits issued as of February 1987
2. Discharges "associated with industrial activity"
3. Discharges "from a municipal separate storm sewer system serving a population of 250,000 or more"
4. Discharges "from a municipal separate storm sewer system serving a population of 100,000 or more but less than 250,000"
5. Other discharges designated by the EPA administrator or the State if such discharge "contributes to a violation of a water quality standard or is a significant contributor of pollutants to waters of the United States"

Final regulations governing stormwater discharges were promulgated in November 1990.

Dredge or Fill Discharge Permit Program

Section 404 of the CWA regulates the discharge of dredged or fill material into waters of the United States. Dredged material is excavated or dredged from a water body. Fill material is that material used to replace water with dry land. The Section 404 permit program is administered by the U.S. Army Corps of Engineers. EPA provides guidelines for the issuance of permits by the Corps of Engineers. States may assume responsibility for portions of the program.

Discharge of Oil and Hazardous Substances

Section 311 of the CWA prohibits discharges of oil or hazardous substances in quantities that may be harmful to waters of the United States. The appropriate Federal agency must be immediately notified of any spill of a "reportable quantity." Section 311 provides for cleanup of spills and requires plans for preparation of Spill Prevention, Control, and Countermeasures (SPCC) plans.

Over 300 substances have been defined as hazardous under Section 311 and each of these substances has a "reportable quantity" (40 CFR, Parts 116 and 117, 1980).

A person or corporation who properly notifies the Agency of the discharge of a reportable quantity of oil or hazardous substance is immune from criminal prosecution but is liable for civil penalties. Additionally, those who cause the spill are liable for the costs of cleanup and removal. If the Federal government must clean up the spill, the discharger of the spill is liable for cleanup costs. There are maximum liability limits depending upon the type of facility and spill. These limits do not apply if the discharge resulted from willful negligence or willful misconduct of the owner.

Certain discharges of oil and hazardous material that flow from a point source may be excluded from Section 311 liability if, during preparation of the NPDES permit covering that facility, conditions are added to the permit to avoid the occurrence of a spill.

Enforcement

Section 309 of the CWA provides several enforcement options which can result in large penalties to violators.

Criminal violations can result in penalties for individuals of up to $250,000 and imprisonment for 15 years for "knowing endangerment", while penalties against organizations for similar violations can reach $1,000,000. "Knowing violations" result in fines of $5,000 to $50,000 per day and imprisonment of up to 3 years <u>per day</u> of violation. "Negligent violations" carry penalties of $2,500 to $25,000 and up to 1 year imprisonment <u>per day</u> of violation. Falsification of reports is punishable by a $10,000 fine and imprisonment of up to 2 years. All penalties may be doubled for second offenses.

Civil penalties may be assessed in an amount up to $25,000 per day of violation. Factors to be considered by the court in determining the amount of a civil penalty include the seriousness of the violation, the economic benefit to the defendant as a result of the violation, compliance history, good-faith efforts applied by the violator and the economic impact of the penalty on the violator.

Administrative penalties may also be imposed against violators through the initiation of an administrative penalty proceeding. Section 309(g) provide for two classes of penalties, Class I and Class II, which differ with respect to the limits on the penalties which can be imposed and the procedures which must be followed in order to impose those penalties.

<u>RESOURCE CONSERVATION AND RECOVERY ACT OF 1976 (RCRA)</u>[*]

RCRA, as enacted in 1976, was designed to establish "cradle-to-grave" control of hazardous wastes by imposing extensive requirements on those who generate and/or handle such wastes. RCRA applies primarily to current activities at active facilities, yet there is authority for addressing imminent hazards and for taking corrective actions based upon past actions. Although RCRA has been amended several times, the most significant amendments are the Hazardous and Solid Waste Amendments (HSWA) of 1984. HSWA requires, among other things, that regulations be promulgated to address underground storage tanks, to establish a schedule for restricting/prohibiting the land disposal of hazardous wastes, and to revamp the toxicity characteristic as a means for determining whether a

---

[*] *43 U.S.C. §§6901 et seq. and Solid Waste Disposal Act amendments of 1980, P.L. 96-482, 94 Stat. 2334.*

waste is hazardous. (The process of promulgating implementing regulations in these areas is ongoing but almost is complete.)

Solid wastes, if land disposed, are regulated through State programs under Subtitle D of RCRA. Hazardous solid wastes are subject to regulation in their generation, transport, treatment, storage, and disposal under Subtitle C of RCRA. Subtitle C of the statute authorizes a comprehensive Federal program to regulate hazardous wastes from generation to ultimate disposal. A waste is hazardous under Subtitle C if it is listed by EPA as hazardous, or it exhibits a hazardous characteristic (corrosivity, reactivity, ignitability, and toxicity) and is not delisted or excluded from regulation. There are special management provisions for hazardous wastes created by small quantity generators and hazardous wastes that are intended to be reused or recycled.

Solid waste includes garbage, refuse and sludge, other solid, liquid, semi-solid, or contained gaseous material which is discarded, has served its intended purpose, or is a mining or manufacturing byproduct. Most industrial and commercial byproducts can qualify as a solid waste. Exclusions from solid waste include domestic sewage, irrigation return flows, materials defined by the Atomic Energy Act, *in situ* mining waste and NPDES point sources.

Solid wastes excluded from regulation as hazardous solid wastes are household waste; crop or animal waste; mining overburden and wastes from processing and benefication of ores and minerals; flyash, bottom ash waste, slag waste and flue gas emission control waste and drilling fluids from energy development. A waste can be "delisted" from the hazardous waste listing or excluded for other reasons. Some materials intended to be reused or recycled are not fully regulated as solid/hazardous wastes, while others, depending upon the type of waste generated and the recycling process used, are fully regulated.

List of Hazardous Wastes

Hazardous waste streams from specific major industry groups and some generic sources (40 CFR, Part 261, Subpart D, §261.31 and 261.32) and well over 200 toxic commercial chemical wastes (i.e., discarded commercial chemical products and chemical intermediates) are included on the list of hazardous waste (40 CFR §261.33). If a commercial chemical substance is on the list, its off-spec species is also considered hazardous when discarded, as are spill residues. Some of the listed wastes are acutely toxic and are more closely regulated than other hazardous wastes [see 40 CFR §§261.33(e), 261.5(e), and 261.7(b)(3)].

Special Management Provisions

- ### Small Quantity Generators

Small quantity generators are those that generate less than 1,000 kg per month of hazardous waste. There are two classes of small quantity generators:

1. Generators of between 100 and 1,000 kg per month that are subject to most of the requirements of 40 CFR Part 262 which apply to fully regulated generators, except that they are allowed to accumulate up to 6,000 kg of hazardous waste and to store waste for up to 180 to 270 days.

2. Generators of less than 100 kg per month that are exempt from regulation under 40 CFR Part 262 so long as they do not accumulate greater than 1,000 kg of hazardous waste, properly identify their wastes, and comply with the less stringent waste treatment, storage and/or disposal requirements of 40 CFR §261.5.

Note that the classification of the generator is a function of the total wastes generated in a calendar month, not each waste stream. In addition, for acutely toxic wastes, if more than 1 kg per month of waste or 100 kg per month of spill residues are generated, all quantities of that waste are fully regulated.

- ## Recycling or Reuse

The type of waste generated and/or the recycling process employed will determine whether recycled/reused materials are a solid/ hazardous waste. Some of these materials are not considered solid wastes, some are solid wastes but not hazardous wastes, while others are hazardous but are not subject to full regulation, and still other of these materials are both solid and hazardous wastes that are fully regulated. The circumstances surrounding the apparent recycling/ reuse of waste materials should be thoroughly documented during inspection.

- ## Land Disposal Restrictions

A major feature of HSWA is the schedule for prohibiting the land disposal of untreated hazardous wastes. The key dates and statutory/regulatory requirements are as follows:

- May 8, 1985 - Landfilling of bulk or noncontainerized liquid hazardous waste or free liquids in hazardous waste is prohibited.

- November 8, 1986 - Land disposal of certain solvents, as well as dioxin containing hazardous wastes (F-series wastes) is prohibited unless treatment standards are met.

- July 8, 1987 - Land disposal of hazardous wastes listed in Section 3004(d)(2) of RCRA (the "California list") is prohibited unless treatment standards are met.

- August 8, 1988 - Land disposal of 1st Third of listed hazardous wastes (primarily F and K wastes) is prohibited unless treatment standards are met.

- June 8, 1989 - Land disposal of 2nd Third of listed hazardous wastes (F, K, U, and P wastes) is prohibited unless treatment standards are met.

- May 8, 1990 - Land disposal of 3rd Third of listed hazardous wastes (the remaining listed wastes) as well as wastes exhibiting hazardous characteristics is prohibited unless treatment standards are met.

Requirements for Generators[*]

- **Identification** - Hazardous wastes must be identified by list, testing, or experience and assigned waste identification numbers.

- **Notification** - No later than 90 days after a hazardous waste is identified or listed in 40 CFR, Part 261, a notification is to be filed with EPA or an authorized State. An EPA identification number must be received.

- **Manifest System** - Implement the manifest system and follow procedures for tracking and reporting shipments. Beginning September 1, 1985, a waste minimization statement is to be signed by the generator [see RCRA Section 3002(b)].

- **Packing** - Implement packaging, labeling, marking, and placarding requirements prescribed by DOT regulations (40 CFR, Parts 172, 173, 178, and 179).

- **Annual Report** - Submittal required March 1 using EPA Form 8700-13.

- **Exception Reports** - When generator does not receive signed copy of manifest from designated TSDF within 45 days, the generator sends

---

[*] *40 CFR Part 262*

Exception Report to EPA including copy of manifest and letter describing efforts made to locate waste and findings.

- <u>Accumulation</u> - When waste is accumulated for less than 90 days, generator shall comply with special requirements including contingency plan, prevention plan, and staff training (40 CFR, Part 265, Subparts C, D, J, and 265.16).

- <u>Permit for Storage More Than 90 Days</u> - If hazardous wastes are retained on-site more than 90 days, generator is subject to all requirements applicable to TSDFs and must obtain a RCRA permit.

<u>Requirements for Transporters</u>[*]

- <u>Notification</u> - No later than 90 days after a hazardous waste is identified or listed in 40 CFR, Part 261, a notification is to be filed with EPA or an authorized State. Receive EPA identification number.

- <u>Manifest System</u> - The transporter must fully implement the manifest system. The transporter signs and dates manifest, returns one copy to generator, assures that manifest accompanies waste, obtains date and signature of TSDF or next receiver and retains one copy of the manifest for himself.

- <u>Delivery to TSDF</u> - The waste is delivered only to designated TSDF or alternate.

- <u>Record Retention</u> - Transporter retains copies of manifest signed by generator, himself, and accepting TSDF or receiver and keeps these records for a minimum of 3 years.

- <u>Discharges</u> - If discharges occur, notice shall be given to National Response Center. Appropriate immediate action shall be taken to

---

[*] *40 CFR Part 263*

protect health and the environment and a written report shall be made to the DOT.

<u>Requirements for Treatment, Storage, or Disposal Facilities (TSDFs)</u>[*]

• <u>Notification</u> - No later than 90 days after a hazardous waste is identified or listed in 40 CFR, Part 261, a notification of hazardous waste management activities is to be filed with EPA or an authorized State by TSDFs, which manage newly identified or listed hazardous waste.

• <u>Interim Status</u> - These facilities include TSDFs; on-site hazardous waste disposal; on-site storage for more than 90 days; in-transit storage for greater than 10 days and the storage of hazardous sludges, listed wastes, or mixtures containing listed wastes intended for reuse. Interim status is achieved by:

    - Notification (see above)

    - Being in existence on November 19, 1980 or on the date of statutory or regulatory changes which require the facility to have a permit

    - Filing a Part A by the date specified in the regulation covering the facility (40 CFR, Parts 261, 264, or 265)

• <u>Interim Status Facility Standards</u> - The following standards and requirements shall be met.

    - General information (Subpart B)
    - Waste analysis plan
    - Security
    - Inspection plan
    - Personnel training

---

[*]   *40 CFR Parts 264 and 265*

- Handling requirements
- Preparedness and prevention
- Contingency planning and emergency procedures (Subparts C and D)
- Records and reports
- Manifest system
- Operating logs
- Annual and other reports (Subpart E)
- Groundwater Monitoring (Subpart F)
- Closure and post-closure plans (Subpart G)
- Financial requirements (Subpart H)
- Containers, tanks, surface impoundments, piles (Subparts I, J, K, L)
- Land treatment, landfills, incinerators, thermal treatment, chemical, physical and biological treatment (Subparts M, N, O, P, Q)
- Underground injection (Subpart R)

- **Permit** - In order to obtain a permit:

  - Facilities with interim status must file a Part B RCRA permit application when directed to do so by EPA or an authorized State and final facility standards must be met or the facility must be on an approved schedule to meet those standards.

  - New facilities and facilities which do not qualify for interim status are to receive a RCRA permit before construction can begin or a hazardous waste can be handled.

State Hazardous Wastes Programs

Under RCRA, states can obtain approval from EPA to implement programs governing hazardous wastes "in lieu of" the federal program administered by EPA. State programs must be "equivalent" to the federal program to obtain approval, and include the ability to enforce program requirements. Once approved the state standards govern all regulated entities and any assessment of a facility's compliance must be based upon those state regulations. Thereafter, when federal standards change, each authorized state must submit a revised program for EPA approval. Until such approval is received, those new standards generally do not have any effect in those states. The major exception to this regulatory scheme is rule-making based upon the HSWA of 1984. HSWA provides that implementing regulations are to take effect at the same time in all states. Authorized states must still modify their programs to include HSWA requirements, but there is no gap in regulation between the time that the Agency promulgates a final HSWA-based rule and the time that the state receives final approval of the program revision which is equivalent to the federal HSWA rule. As a result, until a revised state program addressing all HSWA requirements is approved for an authorized state, the administration and enforcement of the overall hazardous waste program will involve both EPA and the authorized state.

<u>Enforcement</u>

EPA and authorized States may pursue enforcement actions based on administrative orders, as well as judicial actions seeking civil and criminal penalties for RCRA violations.

An administrative action involves issuance of an administrative order requiring compliance with the regulations. Injunctive relief may be sought in a civil action filed in the U.S. District Court. Civil penalties of up to $25,000 per day of violation may be imposed for violations of Subtitle C of RCRA. Failure to comply with an administrative order may result in suspension or revocation of a permit.

Criminal penalties of up to $50,000 and/or 2 years' imprisonment may be imposed for certain "knowing violations." "Knowing endangerment" that places another person in imminent danger of death or serious bodily injury can result in a fine of up to $250,000 and/or 15 years' imprisonment.

COMPREHENSIVE ENVIRONMENTAL RESPONSE, COMPENSATION, AND LIABILITY ACT (SUPERFUND)

The Superfund Act was enacted December 11, 1980. The Federal government is authorized to clean up toxic or hazardous contaminants at closed and abandoned hazardous waste dumps and the government is permitted to recover costs of this cleanup and associated damages by suing the responsible parties involved. Cleanup monies come out of a "superfund" created by taxes on chemicals and hazardous wastes.

The act provides that, when there is a release of hazardous substance, either real or threatened, the parties who operated the vessel or facility which created the release are liable for the containment, removal, remedial action, response, and injury damages to natural resources under Section 107(a). The act also establishes limitations on liability.

If claims are presented to the liable parties but are not satisfied, the act then allows claims to be reimbursed from the Superfund.

Regulatory provisions under Sections 102 and 103 of the act require that release of hazardous substances into the environment be reported unless the release is in accordance with an established permit. Spills of any "reportable quantity," established pursuant to regulations promulgated under the Act, must be reported.

All owners or operators of any facility handling and disposing of hazardous substances or that has handled hazardous substances in the past (including previous owners and operators) were required to inform the EPA Administrator by June 1981 of their facility activities unless they have a RCRA permit or have been accorded "interim status." Failure of notification is a crime and, if the party knowingly fails to provide these data, they are not entitled to the prescribed limits and defenses of liability.

On October 17, 1986, the Superfund Act was amended under the Superfund Amendments and Reauthorization Act (SARA). Those amendments provide mandatory schedules for the completion of various phases of remedial response activities, establish detailed cleanup standards, and generally strengthen existing authority to effect the cleanup of superfund sites.

Enforcement

Civil and criminal penalties and awards are available under CERCLA. Section 106 provides that failure or refusal to comply with an order directing immediate abatement of a release or threatened release of a hazardous substance which creates an imminent and substantial endangerment to the public health or welfare or the environment is punishable by a fine of up to $25,000 per day of violation. Section 109 also provides for penalties of up to $25,000 per day of violation, to be imposed by a U.S. District Court or in an administrative proceeding for failure or refusal to comply with other provisions of CERCLA.

Under Section 109(d), a "bounty" in the amount of up to $10,000 may be paid to any individual who provides information leading to the arrest and conviction of any person for a CERCLA violation.

Criminal penalties of up to $10,000 and imprisonment for 3 years are available under Section 103 for various violations, including failure to notify of a release and falsification of records. Second and subsequent violations may result in imprisonment of up to 5 years.

TOXIC SUBSTANCES CONTROL ACT (TSCA)

TSCA regulates existing and new chemical substances. TSCA applies primarily to manufacturers, distributors, processors, and importers of chemicals. TSCA can be divided into five parts as follows:

Inventory and Premanufacture Notification

EPA has published an inventory of existing chemicals. A substance that is not on this list is considered "new" and requires Premanufacture Notification (PMN) to EPA at least 90 days before the chemical can be manufactured, shipped, or sold (TSCA, Section 5). If EPA does not make a declaration within 90 days to restrict the product, then full marketing can begin and the chemical is added to the inventory. In addition, a manufacturer may obtain a test marketing exemption and distribute the chemical before the 90-day period has expired. Conversely, EPA, in response, may reject PMN for insufficient data; negotiate for suitable data, prohibit manufacture or distribution until risk data are available; or, pending development of a Section 6 rule, completely ban the product from the market or review the product data for an additional 90 days.

Testing

Under TSCA, Section 4, EPA can require product testing of any substance which "may present an unreasonable risk of injury to health or to the environment." Some testing standards are proposed, but no test requirements for specific chemicals are yet in effect.

Reporting and Recordkeeping

TSCA, Section 8(a) deals with general reporting. The "first tier" rule (PAIR) now in effect is a short form seeking production and exposure data on over 2,300 existing chemicals. A "second tier" rule is expected to obtain more detailed data on a relatively small group of chemicals that may become priority candidates for regulation.

Section 8(c) calls for records of significant adverse effects of toxic substances on human health and the environment. It requires that records of alleged adverse reaction be kept for a minimum of 5 years.

Section 8(d) allows EPA to require that manufacturers, processors, and distributors of certain listed chemicals (designated under 40 CFR 716.13) submit to EPA lists of health and safety studies conducted by, known to, or ascertainable by them. Studies include individual files, medical records, daily monitoring reports, etc.

Section 8(e) requires action upon discovery of certain data. Any person who manufactures, processes, or distributes a chemical substance or mixture, or who obtains data which reasonably supports the conclusion that their chemical presents a <u>substantial risk</u> of injury to health or to the environment, is required to notify EPA immediately. Personal liability can only be limited if the company has a response plan in effect.

Regulation Under Section 6

EPA can impose a Section 6 rule if there is reason to believe that the manufacture, processing, distribution or use, or disposal of a chemical substance or mixture causes, or may cause, an unreasonable risk of injury to health or to the environment. Regulatory action can range from labeling requirements to complete prohibition of the product. Section 6 rules are currently in effect for several chemicals including PCBs. A Section 6 rule requires informal rulemaking, a hearing, and a cost-benefit analysis.

Imminent Hazard

This is defined as a chemical substance or mixture causing an imminent and unreasonable risk of serious or widespread injury to health or the environment. When such a condition prevails, EPA is authorized by TSCA, Section 7, to bring action in U.S. District Court. Remedies include seizure of the chemical or other relief including notice of risk to the affected population or recall, replacement, or repurchase of the substance.

Enforcement

Civil penalties may be assessed through administrative proceedings in an amount not to exceed $25,000 per day of violation. Appeals relating to civil penalties are reviewed in the U.S. Court of Appeals.

Criminal penalties for knowing and willful violations of TSCA may be imposed in amounts of not more than $25,000 per day of violation and/or imprisonment for up to 1 year.

Actions to restrain violations, compel compliance, or seize and condemn any substance, mixture, or article may be brought in the U.S. District Courts.

FEDERAL INSECTICIDE, FUNGICIDE, AND RODENTICIDE ACT (FIFRA)

A pesticide is defined as any substance intended to prevent, destroy, repel, or mitigate pests. FIFRA requires registration of all pesticides, restricts use of certain pesticides, authorizes experimental use permits and recommends standards for pesticide applicators and the disposal and transportation of pesticides.

Pesticides are registered for 5 years and classified for either general or restricted usage. Restricted use means that they are to be applied either by or under the direct supervision of a certified applicator. Pesticides must be labeled and specify ingredients, uses, warnings, registration number, and any special use restrictions. Regulations also specify tolerance levels for certain pesticide chemicals in or on agricultural commodities. These limits apply to 310 different compounds and residue tolerances range from 0 to 100 ppm. A few pesticides are also regulated as toxic pollutants under Section 307(a) of the CWA and by Primary Drinking Water Standards under the SDWA.

Enforcement

FIFRA provides for relatively low penalties when compared with many of the other environmental statutes. Civil penalties range from as little as $500 for private applicators on a first offense, to not more than $5,000 per violation for registrants, commercial applicators, wholesalers, dealers, retailers, and distributors. Criminal penalties against private applicators are misdemeanors punishable by fines of not more than $1,000 and/or imprisonment for up to 30 days. Commercial applicators who knowingly violate FIFRA may be fined up to $25,000 and/or imprisoned for up to 1 year. Registrants, applicants for a registration and producers who knowingly violate this statute are subject to fines of up to $50,000 and/or imprisonment for up to 1 year.

Any person who, with intent to defraud, uses or reveals information relating to product formulas acquired pursuant to FIFRA's registration provisions may be fined up to $10,000 and/or imprisoned for up to 3 years.

SAFE DRINKING WATER ACT

The SDWA of 1974 was established to provide safe drinking water to the public. Both primary and secondary drinking water standards have been set by EPA regulations which apply to water after treatment by public drinking water systems. National Interim Primary Drinking Water Regulations were adopted in 1975 to protect public health (40 CFR, Part 141). Regulations covering radionuclides were added in 1976. Regulations for trihalomethanes were promulgated in 1979. Secondary regulations were established in 1979 as guidelines to States to protect the nonhealth-related qualities of drinking water. The 1986 amendments to the SDWA: (1) establish a mandatory schedule, requiring the promulgation of primary drinking water regulations for 83 contaminants, (2) prohibit the use of lead in public water systems, (3) provide civil and criminal penalties for persons who tamper with public water systems, and (4) require closer scrutiny of State programs, including the direct enforcement of drinking water standards, if necessary.

The SDWA also provides for protection of underground sources of drinking water. Final regulations have been issued whereby States are to establish Underground Injection Control (UIC) waste disposal programs to ensure that contaminants in water supplies do not exceed National Drinking Water Standards and to prevent endangerment of any underground source of drinking water. Injection wells are divided into five classes for regulatory handling. Construction and disposal standards are established for the permitting of Class I to III wells. Class I and IV wells are subject to RCRA requirements. Class IV wells are those used by generators of hazardous or radioactive wastes to dispose of hazardous wastes into formations within one-quarter mile of an underground source of drinking water. New Class IV wells are prohibited and existing Class IV wells must be phased out within 6 months after approval or promulgation of a UIC program in the State. There are numerous State regulatory requirements affecting groundwater which should be consulted by multi-media compliance inspectors. In addition, the 1986 amendments to SDWA strengthen EPA's enforcement authority for UIC programs.

<u>Enforcement</u>

Civil penalties of not more than $25,000 per day of violation may be assessed for failure to comply with national primary drinking water regulations [Section 300g - 3(b)], failure of an owner or operator of a public water system to give notice to persons served by it of failure or inability to meet maximum containment level requirements [Section 300(g) - 3(c)], failure to comply with an administrative order requiring compliance [Section 300g - 3(g)], or failure to comply with requirements of an applicable underground injection control program [Section 300h - 2(b)].

Any person who fails or refuses to comply with an administrative order issued where a contaminant is contaminating or is likely to contaminate an underground source of drinking water and may present an imminent and substantial endangerment to human health, is subject to civil penalties of up to $5,000 per day of violation (Section 300i). Administrative orders relating to violations of underground injection regulations may impose penalties of up to $5,000 per day of violation up to a maximum administrative penalty of $125,000.

Tampering with a public water system may result in a criminal fine and/or imprisonment for up to 5 years. Threats to tamper carry fines and/or imprisonment of up to 3 years. Civil penalties may also be imposed against persons who tamper, attempt to tamper, or make threats to tamper with a public water system. The maximum fine is $50,000 for tampering and $20,000 for an attempt or threat to tamper.

EMERGENCY PLANNING AND COMMUNITY RIGHT-TO-KNOW ACT (EPCRA)

EPCRA was enacted as a part of the Superfund Amendments and Reauthorization Act of 1986, as a freestanding provision to address the handling of "extremely hazardous substances" and to establish an extensive information collection system to assist in responding to releases of those substances. EPCRA is comprised of three subtitles: (1) Subtitle A, which establishes the framework for emergency response planning and release notification; (2) Subtitle B, which contains reporting requirements; and (3) Subtitle C, which contains general provisions, including enforcement, penalties, and trade secrets.

Subtitle A - Emergency Planning and Notification

The goal of Subtitle A is to provide States and local communities with the information necessary to adequately respond to unplanned releases of certain hazardous materials. Through the establishment of State Emergency Response Commissions (SERC) and Local Emergency Planning Committees (LEPC), Subtitle A mandated the development and implementation of emergency response plans. Subtitle A also requires facilities at which certain "extremely hazardous substances" are present in excess of established threshold planning quantities to notify the State Commission of the presence of the substances and to report releases of those substances in excess of specified reportable quantities.

Subtitle B - Reporting Requirements

Sections 311 - 313 of EPCRA (Subtitle B) contain reporting requirements for facilities at which "hazardous chemicals" are present in excess of specified thresholds or which experience environmental releases of "toxic chemicals" in excess of the established threshold quantities.

Section 311 requires facilities at which "hazardous chemicals" are present in amounts exceeding threshold levels, to submit material safety data sheets (MSDSs) or lists of substances for which they maintain MSDSs to the SERC, LEPC, and local fire departments in order to give notice to those authorities of the types of potential hazards present at each facility.

Section 312 requires submission of annual and daily inventory information on the quantities and locations of the hazardous chemicals. "Tier I" reports provide the required general information. "Tier II" reports providing chemical-specific information must be submitted in place of Tier I reports upon request of the SERC, LEPC, or local fire department.

Section 313 requires annual reporting to EPA and the State of any environmental releases of listed "toxic chemicals" in excess of specified threshold quantities. A facility is required to submit a "Form R" Toxic Chemical Release Inventory Report in the event of a release if it has 10 or more full-time employees; is grouped in SIC codes 20 through 39; and manufactures, processes, or otherwise uses a toxic chemical in excess of the established reporting thresholds.

Enforcement

Section 325 of EPCRA sets forth the civil, criminal, and administrative penalties which may be assessed for violations of that Act. Violation of an administrative order may result in civil penalties of up to $25,000 per day. Penalties for violations of the emergency notification provisions of Section 304 may be assessed through administrative or judicial proceedings, with potential penalties ranging from $25,000 per violation to $25,000 per day of violation. Any person who knowingly or willfully fails to provide emergency notification may be assessed a criminal penalty of up to $25,000 and/or 2 years' imprisonment, ($50,000 and/or 5 years for second and subsequent convictions).

Violations of reporting requirements carry civil penalties of up to $25,000 per violation. Frivolous trade secret claims may result in penalties of up to $25,000 per claim, whereas the knowing and willful disclosure of actual trade secret information may be punishable by a fine of up to $20,000 and/or imprisonment up to 1 year.

APPENDIX B

MULTI-MEDIA INSPECTIONS
DEFINITIONS AND TRAINING

MULTIMEDIA INSPECTIONS – DEFINITIONS AND TRAINING

06-05-91

To describe a proposed program for multimedia inspections and inspector training, we must first establish some definitions:

1. Compliance Programs – These are administrative, civil, and criminal enforcement programs as authorized to EPA.

2. Inspection categories – All inspections, for purposes of this program description can be grouped into four categories.

Category A – An inspection for a single program. A program may have one or more types of inspections.

Category B – A simplified screening multimedia inspection is conducted in addition to the single program (Category A) inspection. The screening inspection is intended to identify the more obvious and readily detectable instances of non-compliance or indicators of possible non-compliance in other compliance program areas with only minimal additional resources required. Information obtained would be referred to the appropriate compliance program office(s) for follow-up. Follow-up action could include a full inspection for one or more programs or in some instances immediate enforcement action. The inspector would use a simplified checklist as a guide and to record observations and pertinent information. As an example, observations of potential compliance issues could include inoperable control systems, unusual emissions/ discharges, evidence of spillage or leaks, breached dikes, new emission/discharge sources, lack of permits or SPCC plan, abandoned drums, etc.

Category C – A multi-program inspection is an inspection of a facility for two or more (Category A inspections) compliance programs but not compliance with all applicable requirements. The inspection may be conducted by either one or a team of inspectors. The team which is headed by a team leader, conducts a detailed inspection for each of the targeted programs represented. The inspection may include screening for the more obvious and readily detectable indicators of possible non-compliance in other compliance program areas.

Category D – A multimedia inspection is a comprehensive inspection which addresses relevant program compliance requirements operative for a single facility at a specific point in time that can be used in subsequent enforcement actions. The inspection is usually conducted by a team of inspectors lead by a team leader. Team members may be comprised of ESD, program,

Page 2

contractors, state, NEIC, or local agency inspectors. The inspection may focus on facility processes to identify activities and wastes potentially subject to regulations and to identify potential cross-program issues.

3. Inspector Training - For purposes of this program description, inspector training will be described in terms of four inspector training levels.

Level 1 -   The Level 1 (single program) inspector is a person who has been trained to be a lead inspector per EPA Order 3500.1 (Basic Health & Safety, Fundamentals curriculum, and program specific minimum) for a single compliance program.

Level 2 -   The Level 2 (screening) inspector is a person who has received training beyond Level 1 to screen for and report on the more obvious or key indicators of non-compliance in all environmental program areas relevant to a particular facility or site (Category B inspection). The additional training requires a minimum of time (1 to 2 days) is keyed to a simplified checklist, and is designed to enable the inspector to ask key questions and readily recognize the more obvious environmental problems or key indicators of non-compliance. Additionally, the training will enable the inspector to readily recognize and report the more obvious violations of OSHA regulations and as a minimum disseminate pollution prevention information. The training could be heavily aided by extensive use of videos to demonstrate critical observations.

Level 3 -   The Level 3 (multi-program) inspector is a person who has been trained to be a lead inspector per EPA Order 3500.1 for two or more compliance programs and has started the leadership project management and team building training required for Level 4 inspector training. Level 3 inspector training is a prerequisite for Level 4.

Level 4 -   The Level 4 (multimedia) inspector is a senior, experienced person who has received training beyond Levels 2 or 3 to lead a team of inspectors to conduct either a multimedia (Category D) inspection which addresses all relevant compliance program requirements or a multi-program (Category C) inspection which addresses the requirements of two or more compliance programs. The inspector in addition to having significant experience in leading inspections, should complete the following:

Page 3

> * leadership project management and team building
>   training
> * training to recognize cross-media impacts and
>   integrate cross media evaluations
> * training and experience to merge diverse written
>   reports into a coherent whole.

A summary in matrix form, of multimedia inspection and inspector training definitions are shown in Figure 1.

PROGRAM GOALS

The overall goal of each region would be a multimedia compliance program with
inspectors in each region equipped to perform all four categories of inspections.
The specific goals of the FBCs is:

1. To have all inspectors trained per EPA Order 3500.1 (Level 1) to as a
    minimum, be able to conduct inspections for a single compliance
    program (Category A).

2. To have all inspectors trained to be Level 2 inspectors and conduct
    screening (Category B) inspections during all compliance inspections by
    the end of FY 1993.

3. To have teams of dedicated multi-program (Level 3) and multimedia (Level
    4) inspectors trained to conduct or lead multi-program (Category C) or
    multimedia inspections (Category D) as required by the end of FY 1993.

4. To have inspectors progress on a career ladder from Level 1 to Level 4
    inspectors.

5. To build into the inspections and inspector training program as
    appropriate the reporting of enviromental results.

These goals require a dedicated effort to develop and improve upon the skills
that would be required for a single program inspector. Regional compliance
programs can be agressive in targeting multi-program and multimedia inspections
to achieve environmental results using TRI data, comparative risk, geographic,
and other enforcement initiatives. Similarly, the FBCs can take the initiative
to identfy overlapping routine program inspection requests and flag potential
multi-program enforcement opportunities. Achievements of these goals would
provide enhanced capability for responding to special requests from the RA/DRA
and providing support to the office of criminal investigations.

MULTIMEDIA INSPECTIONS

DEFINITION AND TRAINING NEEDS

INSPECTION CATAGORY / INSPECTOR TRAINING LEVEL	A Single Program Inspection	B Screening Inspection *(Single Program plus Screening)*	C Multi Program Inspection *(2 or more Programs)*	D Multimedia Inspections *(All Relevant Requirements)*
1. Lead in one program.	Lead		Team Member	Team Member
2. Level 1 Plus Screening Training	Lead	Lead	Team Member	Team Member
3. Lead in Two or More Programs	Lead		Lead/Team Member	Team Member
4. Level 3 Plus Team Leader Training	Lead		Lead	Lead
Non Lead Inspector	Team Member	Team Member	Team Member	Team Member

APPENDIX C

TEAM LEADER RESPONSIBILITIES AND AUTHORITIES

Appendix C
TEAM LEADER RESPONSIBILITIES AND AUTHORITIES

<u>INTRODUCTION</u>

The team leader is the lead person for a given project. The team leader for each project is selected by the Branch Chiefs, based on factors such as project needs, employee development opportunities, and personnel availability. In general, the team leader is a work group leader, the central focal point for a particular project, responsible for ensuring that project objectives are met in a timely manner. The team leader is given certain responsibilities and authorities, as outlined below, and is expected to fulfill the responsibilities and exercise authority to successfully plan, coordinate, conduct, and complete the project.

The extent of input and involvement by the supervisor in these team leader responsibilities is dependent on the team leader's grade and experience. GS-12 and GS-13 team leaders should be able to perform most, if not all, required tasks. Team leaders of other grades will require more assistance from their supervisor or mentor.

Team leader responsibilities and authorities may vary somewhat with unique requirements of each project. However, general responsibilities and authorities for conducting a complete and timely project are common for most projects. The following discussion of team leader responsibilities is presented in two sections: responsibilities and authorities. The discussion of responsibilities is presented by project phases (most projects will involve some form of each project phase). The discussion of authorities follows. Because the team leader has similar authorities for most project phases, the authorities are not discussed in terms of project phases.

TEAM LEADER RESPONSIBILITIES

Phase 1 - Project Request/Project Objectives

As stated previously, the team leader is the central focal point for a given project responsible for assuring that project objectives are met in a professional and timely fashion. The team leader uses the media specific and Multi-Media Investigation manuals as guidance for conducting environmental compliance investigations. General project phases and associated team leader responsibilities follow. Phase 1 of any project begins with a request for assistance. Depending on the specifics of the request, work is required to develop that request into a project plan that addresses the requestor's needs.

The team leader, in conjunction with his/her supervisor, project requestor and, often times, members of the project team, is responsible for:

- Defining project objectives

- Defining specific tasks required to fulfill project objectives

- Identifying resource needs (both equipment and personnel)

- Identifying potential on-the-job training (OJT) opportunities associated with the various project phases and, in conjunction with supervisors (and, in some cases, staff), develop OJT objectives for other personnel

- Scheduling project tasks

- Coordinating with supervisors to ensure availability of project members for project tasks

- Developing and assigning work tasks to team members (this should include identifying any OJT opportunities)

- Ensuring that a comprehensive Project Plan is prepared (this task may be a separate project phase - project phase 3 - depending on the extent of information available during the project request)

- Maintaining communication with project requestor and appropriate personnel (team members, team member supervisors, counterparts in other agencies, etc.)

Phase 2 - Background Information Retrieval and Review

This project phase involves identifying, collecting and reviewing background information applicable to a specific project. The team leader, often in conjunction with project team members, is responsible for:

- Identifying necessary background information (including applicable laws and regulations, facility description, past compliance status, safety considerations, etc.)

- Arranging for access to background information

- Arranging for receipt of background information (such as file review and document retrieval, phone requests, phone conversations, etc.)

- Ensuring that background information is properly organized and filed

- Ensuring that background information is provided to appropriate project personnel

- Conducting reconnaissance inspection, if appropriate

- Providing guidance to project members and ensuring that any associated OJT objectives are met

- Maintaining lines of communication with appropriate NEIC personnel (team members, team member supervisors, etc.)

Phase 3 - Project Plan

The project plan is a written document completed at some point prior to the field work on most projects. The plan identifies work to be conducted to address project objectives and includes a site safety plan. The team leader is responsible for:

- Overall completion of a final project plan (including site safety plan, if site work is required)

- Ensuring that the project plan is peer reviewed

- Obtaining concurrence from the project requestor

- Providing all project team members with copies of the project plan

- Ensuring that all project members are familiar with the contents of the plan, including individual project responsibilities, project schedules, and safety requirements

- Ensuring that the project requestor receives the final project plan prior to any on-site work

- Providing guidance to project members and ensuring that any associated OJT objectives are met

Phase 4 - On-site Investigation

This project phase involves the on-site field work and necessary logistics/personnel actions to ensure that the field investigation is carried out in a complete, efficient, and timely manner. During this project phase, the team leader is responsible for:

- Ensuring that all personnel actions (overtime/compensatory time, work schedule changes, etc. are addressed)

- Ensuring that logistical issues are addressed (transportation of personnel and equipment to the site, lodging arrangements, etc.)

- Developing and maintaining a working relationship between all parties involved (including investigation target and contractors)

- Coordinating all on-site activities, including scheduling

- Ensuring that all project objectives are addressed during the on-site investigation

- Ensuring that the site safety plan is followed (or more stringent facility requirements, if appropriate)

- Maintaining communication with appropriate personnel (team members, team member supervisors, etc.) and with other appropriate personnel (such as project requestor, contractors, Department of Justice, FBI)

- Providing guidance to team members and ensuring that any associated OJT objectives are met

- Ensuring that contaminated equipment is properly disposed of or cleaned

- Directing public inquiries to proper authorities

Phase 5 - Information Evaluation/Report Preparation

A report of project activities, information evaluation, and findings is prepared for most projects. The project report is usually the major conduit for presenting project findings to the project requestor. In general, the coordinator is responsible for ensuring that the project report addresses all the project objectives, is accurate, and is reviewed and completed in a timely fashion.

- Preparing a report outline (or otherwise identifying report structure and contents to project members)

- Identifying and assigning preparation of project report sections to individual project members

- Identifying and communicating report writing schedules to project members

- Coordinating with supervisors to ensure availability of project members for report preparation

- Maintaining communication with project requestor and appropriate personnel (team members, team member supervisors, etc.)

- Providing guidance to project members and ensuring that all associated OJT objectives are met

- Coordinating all portions of report preparation with other groups (Graphics, Report Services, etc.)

- Assembling draft reports

- Ensuring that report is properly reviewed and revised (including transmittal of drafts for external review, tracking of copies, and return of external draft copies)

- Assembling and transmitting final report

<u>Phase 6 - Project Follow-up</u>

Project follow-up includes project activities that follow transmittal of the final report. This normally includes providing input into legal action such as case preparation, court testimony, settlement negotiations, and depositions. During project follow-up, the team leader is responsible for:

- Maintaining post-project report contact with project requestor or other designated project contact to remain informed of legal or other activities

- Coordinating any requests for additional assistance

- Advising supervisors and appropriate staff of additional or potential additional project support

- Preparing project file for turn-in to central files

- Providing the supervisor of each project member a critique of the project member's activities, including a discussion of the OJT objectives as identified by project coordinator and project member's supervisor during project request phase

- Provide each team member with an individual, verbal critique of his/her performance

TEAM LEADER AUTHORITIES

As the central focal point and leader of a team of employees, the team leader has some "first-line" supervisor authorities for most project phases. Once project members are selected and general responsibilities of each member are agreed upon (agreement between team leader and project member's supervisor and often times the project member), the team leader has the authority to:

- Set and/or modify project schedules

- Identify and modify, as necessary, specific project staff tasks (including activities in all project phases such as project plan preparation, field work, and report preparation).

- Direct field operations

- Enforce project safety plan requirements (including barring personnel without the proper equipment and/or training from the identified "hot" zone)

- Set working hours for team members during field work

- Approve/verify OT/CT hours worked

- Negotiate terms of inspection with company (e.g., taking of photographs, requesting/copying documents, advance notification of areas to be inspected, personnel to be interviewed, handling of CBI materials, etc.) so long as no statutory or regulatory authorizations are compromised

- Arrange for inspection of off-site facilities related to project objective (e.g., off-site contractor laboratory, waste transfer station, etc.)

- Request/arrange for assistance from other groups (e.g., Laboratory Services, ORD, EPIC, etc.)

- Procure project-specific equipment/services

- Require project team members to follow established protocols for conduct, investigation, preparing reports, and participation in follow-up actions, and enforce ground rules identified for specific investigations

- Ensure security of project files report, etc. and investigation findings

APPENDIX D

TYPES OF INFORMATION

Appendix D
TYPES OF INFORMATION

<u>TECHNICAL INFORMATION</u>

<u>Facility Background</u>

- Maps showing facility location and environmental and geographic features (stacks, discharge pipes, and solid waste disposal sites)

- Geology/hydrogeology of the area

- Aerial photographs

- Names, titles, phone numbers of responsible facility officials

- Process description, process flow charts, and major production areas

- Records reflecting changes in facility conditions since previous audit/permit application

- Production levels - past, present, and future

<u>Audit Reports, Records, and Files</u>

- Federal and State compliance files

- Correspondence between the facility and the local, State, and Federal agencies

- Citizens' complaints and reports, follow-up studies, findings

- Audit records, reports, correspondence on past incidents or violations

- Emissions inventory

- Self-monitoring data and reports

- EPA, State, and consultant studies and reports

- Annual reports by the facility (e.g., PCB annual documents and inventories, Securities Exchange Commission §10K reports)

- Records, applications, reports, manifest files, etc. (e.g., RCRA reports, CERCLA submittals)

- Laboratory audit reports, QA/QC activities

- Records of previous hazardous substances spills and malfunctions

Pollutant and Waste Generation, Control, Treatment, and Disposal Systems

- Description and design data for pollution control systems and process operations

- Sources and characterization of wastewater discharges, hazardous wastes, emissions, types of treatment, and disposal operations

- Type and amount of waste generated which is discharged, emitted, stored, treated, and disposed

- Waste storage, treatment, and disposal areas

- Waste/spill contingency plans

- Available bypasses, diversions, and spill containment facilities

- Industrial process, pollution control, treatment and disposal methods, monitoring systems

<u>Legal Information</u>

<u>Requirements, Regulations, and Limitations</u>

- Permit applications, draft or existing permits, registrations, approvals, and applicable Federal, State, and local regulations and requirements

- Application certificates, EPA identification numbers

- Information on draft permits which is different from current conditions

- Exemptions and waivers

- Receiving stream water quality standards, ambient air standards, State Implementation Plans, protected uses

- RCRA notification and Part A and Part B applications

- Pesticide labels

- Grant applications for publicly owned treatment works, research and development demonstration projects and progress reports on these projects

- Federal and State classification of facility (e.g., Interim Status, Small Quantity Generator)

D-4

<u>Enforcement History</u>

- Status of current and pending litigation against the facility[*]

- Deficiency notices issued to facility and responses by the facility

- Status of administrative orders, consent decrees, and other regulatory corrective actions, if any, and compliance by the facility

- Penalties imposed against the company

<u>Information Sources</u>

<u>Laws and Regulations</u> - Federal laws and regulations establish procedures, controls, and other requirements applicable to a facility [Table 1]. In addition, State laws and regulations and sometimes even local ordinances may be applicable, or take precedence.

<u>Permits and Permit Applications</u> - Permits provide information on the limitations, requirements, and restrictions applicable to discharges, emissions and disposal practices; compliance schedules; and monitoring, analytical, and reporting requirements. Applications provide technical information on facility size, layout, and location of pollution sources; waste and pollutant generation, treatment, control and disposal practices; contingency plans and emergency procedures; and pollutant characterization - types, amounts, and locations of discharge, emissions, or disposal.

<u>Regional and State Files</u> - These files often contain grant records, applications, facility self-monitoring data, and audit reports, as well as permits and permit applications pertaining to individual facilities.

---

[*] *Coordination should occur prior to the audit (in conjunction with the EPA Regional Office) with the local Assistant United States Attorney or Justice Department attorney responsible for the civil or criminal case and any consent decree.*

Table 1

FEDERAL STATUTES/REGULATIONS FOR MULTI-MEDIA INVESTIGATIONS

	Air CAA	Water CWA	Superfund CERCLA and EPCRA	Pesticides FIFRA	Solid Waste RCRA	Drinking Water SDWA	Toxics TSCA
Inspection Authority	114,[a] 211[a] [80, 86[b]]	308, 402 [122.41]	104	8, 9 [160.15, 169.3]	3007, 9005 [270.30]	1445 [142.34, 144,51]	11 [717.17, 792.15]
Recordkeeping Authority	114, 208 [51, 57, 58, 60, 61, 79, 85, 86]	308, 402 [122.41, 122.48]	103 372.10	4, 8 [160.63, 169, 160.185-195]	3001, 3002, 3003, 3004, 9003 [262.40, 263.22, 264.74, 264.279, 264.309, 265.74, 264.309, 270.30]	1445 [141.31-33, 144.51, 144.54]	8 [704, 710, 717.15, 720.78, 761.180, 762.80, 763.114, 792.185-195]
Confidential Information	208, 307 [2.201-2.215, 2.301, 53, 57, 80]	308 [2.201-2.215, 2.302, 122.7]	104 [2.201-2.215] 322 [350]	7, 10 [2.201-2.215 2.307]	3007, 9005 [2.201-2.215, 2.305, 260.2, 270.12]	1445 [2.201-2.215, 2.304, 144.5]	14 [2.201-2.215, 2.306, 704.7, 707.75, 710.7, 712.15, 717.19, 720.80-95, 750.16, 750.36, 762.60, 763.74]
Emergency Authority	303	504	104, 106 [300.53, 300.65]	27 [164.166]	7003	1431	7
Employee Protection	322	507	110		7001	1450	23
Permits Basic requirements include applications, standard permit conditions, monitoring, reporting		[122, 125]			[270]	[144, 147]	
EPA procedures for permit issuance	[124]	124]			[124]	[124]	
Technical requirements	[52]	[129, 133, 136, 302[c] BMP[d] [125] SPCC[e] [112] Waivers [125, 130]			[260-266]	[146, 264]	
Specific References	NSPS[f] NESHAP[g] [61] CEM[h] [60] SIP[j] [52] PSD[k] [50]	Effluent guidelines [400-460] BMP [125], SPCC [112], Pretreatment [125, 403], Toxic [129]			Generators [262], Transporters [263], TSD[i] [265], Stds. for TSD Permits [264], Interim Stds [265], Storage <90 days [262], Exemptions [261]		PCBs [761] Dioxin [775]

a   Statute (e.g., Clean Air Act, Section 114 or 211)
b   [80, 86] - 40 CFR, Parts 80 and 86; CFR refers to Code of Federal Regulations
c   Reportable quantities
d   BMP - Best Management Practices
e   SPCC - Spill Prevention Control and Countermeasures Plan

f   NSPS - New Source Performance Standards
g   NESHAP - National Emission Standards for Hazardous Air Pollutants
h   CEM - Continuous Emission Monitoring
i   TSD - Treatment, Storage and Disposal
j   SIP - State Implementation Plan
k   PSD - Prevention of Significant Deterioration

These information sources can provide compliance, enforcement, and litigation history; special exemptions and waivers applied for and granted or denied; citizen complaints and action taken; process operating problems/solutions; pollution problems/solutions; and laboratory capabilities. Consultant reports can provide design and operating data and recommendations for processes; pollutant sources; treatment, control, and disposal systems; and remedial measures.

<u>Technical Reports, Documents, and References</u> - These sources provide information on industrial process operations, data on available treatment, control and disposal techniques, such as their advantages or drawbacks, limits of application, etc. Such sources include Effluent Guideline and New Source Performance Standard development documents and EPA's Treatability Manual. Similar guidance documents on hazardous waste generation, treatment/ disposal are also available.

The background information sources for overall program areas and those that apply specifically to the water, air, solid waste, pesticides, and toxic substances programs are listed in Table 2.

Table 2

BACKGROUND REVIEW INFORMATION SOURCES

Overall Program Areas	Water Pollution CWA	Air Pollution CAA	Solid Wastes Pollution RCRA	Toxic Substances Pollution TSCA	FIFRA/CERCLA
NEIC Information Retrieval System data on corporate structure, financial conditions, pollution control history, environmental and health impacts of pollutants of interest.	NPDES permits/permit applications/draft permits	Air permits and permit applications (Federal/State/local)	Part A of permit application (TSDs only) to designate type and volume of wastes handled, type and design capacity of treatment, storage and/or disposal processes	Available information on chemical substances produced by facility	FIFRA Establishment numbers, Certified Applicator numbers
EPA grants (R&D, constructing, planning)	Applicable effluent guidelines	Self-monitoring requirements and self-reported data		Applicable regulations regarding manufacture, identification, self-reporting requirements, concerning toxic materials (e.g., PCB rules)	Applicable labels
Information available on process operations; pollutants of interest; existing treatment, control and disposal practices; raw material	Compliance inspection reports (Federal/State/local)	Compliance inspection reports (Federal/State/local)	Part B of permit application, if available		Inspection reports (Federal/State)
	Laboratory performance reports	Applicable NESHAP	Draft/final RCRA permit	Inspection reports (Federal/State/local)	EPA Pesticide Inspection Manual
Administrative Orders issued for environmental noncompliance	Self-monitoring requirements and self-reporting data	Applicable NSPS	Applicable regulations for source designations	Technical manuals and references on applicable treatment/control and disposal technology, inspection and monitoring procedures and techniques	State Facility Permits for procedures, bulk storage
Applicable local ordinances on environmental control	Best Management Practices Plan	Applicable air quality standards	Groundwater monitoring plans/data		
Compliance history and present compliance status	Spill Prevention Control and Countermeasure Plan	State Implementation Plan	UIC permit and present status		CERCLA Preliminary Assessment (PA) reports
Available correspondence between regulating officer and facility officials	Pretreatment requirements if facility discharges to POTW	Ambient air quality reports for AQCR	Hydrogeologic reports on local area relative to UIC permit		Site Inspection (SI) reports
Available contractor/ consultant report on facility environmental control matters	Applicable Federal/State regulations related to water pollution control at facility	Stack test reports	Self-monitoring requirements and self-reported data		Remedial Investigation/Feasibility Study (RI/FS) reports
Environmental compliance schedules and present status	Technical manuals and references on pollution treatment/control technology, process operation, monitoring inspection procedures	Air pollutants emission inventories	Applicable regulation on manifest requirements		Records of Decision (RODs)
Available aerial photography	Interstate Commission water quality data (Ohio River Sanitary Commission, Delaware River Basin Commission, Interstate Commission on the Potomac River	Continuous monitoring practices and facility and applicable performance inspections	Inspection reports (Federal/State/local)		Remedial Design (RD) reports
		Available contractor/consultant reports	Technical manual and references on applicable treatment/control and disposal technology, inspection and monitoring procedures and techniques		Removal Action reports
		Technical manuals and references on applicable pollution treatment/control technology, process operators, air pollution monitoring, inspection procedures			

APPENDIX E

SOURCES OF INFORMATION

<u>APPENDIX E</u>

SOURCES OF INFORMATION

I.   General References:

A.   NEIC Policies and Procedures Manual - Covers chain-of-custody, shipping, document handling, report preparation, and in general, how to conduct an investigation [EPA 33019-78-001-R, August 1991].

B.   RCRA Orientation Manual, 1990 Edition. USEPA, Office of Solid Waste/Permits and State Programs Division and the Association of State and Territorial Solid Waste Management Officials. GPO 1990-261-069/24136H

C.   Standard Operating Safety guides. USEPA Office of Emergency and Remedial Response, Emergency Response Division. July 1988. GPO 1988-548-158/87012.

D.   EPA Publications Bibliography. Quarterly listing of all EPA publications distributed through the National Technical Information Service, indexed alphabetically, numerically, and by key word. NTIS, U.S. Department of Commerce, Springfield, VA 22161 (703) 487-4650.

E.   Access EPA: Libraries and Information Services. NTIS, U.S. Department of Commerce, Springfield, VA 22161 (703) 487-4650.

B.   Computer Data Systems - A description of the automated data systems accessed by NEIC. Indexes 41 sources accessing over 1,000 data bases.

II.   Technical References:

A.   Kirk-Othmer Encyclopedia of Chemical Technology. Wiley, 3rd ed., 1981; 4th ed. in publication process.

B.   Merck Index: Encyclopedia of chemicals, drugs, and biological compounds. Good source for chemical properties and safety plan details.

C.   Directory of Chemical Producers: Lists major chemical producers and the products they make. SRI, International: Menlo Park, California. Annual.

III. Legal/Regulatory References:

    A. Statutes at Large: The official publication of a public and private laws and resolutions enacted during a session of Congress.

    B. United States Code: A codification of the general and permanent laws of the United States. New editions appear approximately every 6 years with cumulative annual supplements.

    C. Regulations

        1. Federal Register. Daily publication of proposed and final rules.

        2. Code of Federal Regulations: Annual compilation of regulations.

        3. LSA (Lists of CFR Sections Affected): Monthly updates of CFR by section.

IV. Computer Data Systems - A description of the automated data systems accessed by NEIC follows:

CONTENTS

AGENCY INTERNAL INFORMATION SYSTEMS CURRENTLY ACCESSIBLE BY NEIC

Aerometric Information Retrieval System (AIRS)
AIRS Facility Subsystem (AFS)
Chemicals in Commerce Information System (CICIS)
Comprehensive Environmental Response, Compensation and Liability Information System (CERCLIS)
Consent Decree Tracking System (CDETS)
Docket System
DUNS Market Identifiers (DMI)
Emergency Response Notification System (ERNS)
Enforcement Document Retrieval System (EDRS)
Facility and Company Tracking System (FACTS)
Facility Index System (FINDS)
Federal Reporting Data System (FRDS)
NPDES Industrial Permit Ranking System
Permit Compliance System (PCS)
Pollution Prevention Information Exchange System (PIES)
Potentially Responsible Parties System
Records of Decision System (RODS)
Resource Conservation and Recovery Information System (RCRIS)
Site Enforcement Tracking System (SETS)
STORET
Superfund Financial Assessment System (SFFAS)
TECHLAW Evidence Audit System
Toxic Release Inventory System (TRIS)

AGENCY INFORMATION SYSTEMS NOT CURRENTLY ACCESSIBLE BY NEIC

FIFRA/TSCA Tracking System (FTTS)

CONTENTS (cont.)

PUBLICLY AVAILABLE EXTERNAL INFORMATION SYSTEMS CURRENTLY ACCESSIBLE BY NEIC

Bibliographic Retrieval Service (BRS)
Chemical Information System (CIS)
Colorado Alliance of Research Libraries (CARL)
DataTimes
DIALOG Information Services, Inc.
Dun and Bradstreet
Groundwater On-Line (GWOL)
Justice Retrieval and Inquiry System (JURIS)
NEXIS/LEXIS
National Library of Medicine (NLM)
Scientific and Technical Information Network (STN)
VU/TEXT
WESTLAW

AGENCY INTERNAL INFORMATION SYSTEMS
CURRENTLY ACCESSIBLE BY NEIC

System	Description	Application
Aerometric Information Retrieval System (AIRS)	A national system in ADABAS maintained by the National Air Data Branch which incorporates information from many of the Agency's air databases. Emissions data (formerly in NEDS) is now available in AIRS.	Data currently available from AIRS consists of the ambient air quality data collected by States, utilized for trends analysis and pollution control strategies and emissions and compliance data collected by EPA and State agencies.
Chemicals in Commerce Information System (CICIS)	A national system containing the results of the 1977 TSCA inventory and later cumulative supplement of approximately 60,000 unique chemical substances (7,000 claim confidentiality) used commercially in the United States.	NEIC can access the system by company name and geographical area, generate listings by company name, CAS registry number or geographical area.
AIRS Facility Subsystem (AFS)	A national system containing compliance information including compliance status, agency actions (e.g., inspections), etc. for major sources of the five primary air pollutants. Recently converted from the Compliance Data System (CDS), AFS is one of five AIRS subsystems.	NEIC can acquire the Significant Violators list and compliance event data for individual sources, whole facilities, sources within a certain geographical area and sources of a specific industrial classification.
Comprehensive Environmental Response, Compensation and Liability Information System (CERCLIS)	A national system containing names and locations of uncontrolled hazardous waste sites in the U.S., summary response event status information, alias names and site characteristic data. Recent modifications include provisions for tracking enforcement activities, and technical and chemical information at CERCLA sites. Superfund Comprehensive Accomplishment Plan (SCAP) data is also available through CERCLIS.	NEIC can generate site inventory listings for geographical area, the National Priorities List, technical event status reports, and enforcement history for any uncontrolled hazardous waste site and cleanup expenditure reports.
Consent Decree Tracking System	A national system containing a computerized inventory of consent decrees to which EPA is a party, and computerized summaries of the contents of decrees by facility. NEIC maintains a hardcopy library of all consent decrees within the system. This repository has been converted to a full-text database on JURIS.	NEIC can produce hard copies of all decrees in the inventory, and produce computer reports of the inventory, the entire contents of decrees, the milestones to be met in specific decrees or for decrees within a Region and the contents of all decrees for a specific issue (e.g., groundwater monitoring).

AGENCY INTERNAL INFORMATION SYSTEMS
CURRENTLY ACCESSIBLE BY NEIC

System	Description	Application
Docket System	A national system containing all pertinent information regarding a civil or administrative enforcement action taken by EPA or designated States against violators of all Federal environmental statutes.	NEIC can access the entire system to produce reports of enforcement actions in a geographical area, for a specific statute or media or for a specific source classification.
DUNS Market Identifiers (DMI)	Leased by the Agency from Dun and Bradstreet, DMI contains basic business information for privately- and publicly-owned companies in the United States.	NEIC can generate reports with business information such as number of employees, amount of sales, telephone number, principal officer title, and line of business.
Emergency Response Notification System (ERNS)	A national system containing information on reported releases of oil and hazardous substances and responses by EPA, the U.S. Coast Guard and others to the reported releases.	Reports can be generated to identify specific releases and to aggregate data on the number and types of releases throughout the country and in specific states and regions.
Enforcement Document Retrieval System (EDRS)	EDRS is a full-text national database of EPA enforcement documents including the General Enforcement Policy Compendium and the Policy Compendiums for FIFRA, TSCA, RCRA, CERCLA/ SARA and CWA/FWPCA.	EDRS can be used to retrieve all enforcement documents containing a word like "landfill" or relevant to an issue, law or regulation.
Facility and Company Tracking System (FACTS)	A national database which provides basic business information for privately- and publicly-owned companies in the United States, and facility information for EPA-regulated facilities. FACTS is comprised of the DMI and FINDS subsystems.	NEIC can generate facility listings for any geographical area, type of business, and/or corporation. DMI locates business information such as number of employees, amount of sales, telephone number, and principal officer. FINDS provides facility information for EPA-regulated databases.
Facility Index System (FINDS)	A national database which serves as a cross-reference index on a facility basis to point to media-specific EPA databases to acquire additional data. This is the link with other EPA data systems.	NEIC can generate facility listings for any geographical area, as well as a tabulated listing of whether other databases contain information about that facility.

AGENCY INTERNAL INFORMATION SYSTEMS
CURRENTLY ACCESSIBLE BY NEIC

System	Description	Application
Federal Reporting Data System (FRDS)	A national system containing an inventory of public water supplies in support of the Safe Drinking Water Act. It contains identification and statistical summary information for each public water supply including type of data collected or monitored, and analytical procedures.	NEIC can acquire source information and location, service areas, geographic areas, and historical information. Information on noncompliance and enforcement actions can also be obtained.
NPDES Industrial Permit Ranking System	An NEIC-operated and maintained system which contains criteria, ranking factors and calculation mechanisms to rate (1) a facility's effluent discharge pollution potential, including toxics; (2) health impact potential; and (3) water quality impact potential which is then used in PCS for major/minor differentiation.	NEIC can access the specific data for any of 12 criteria, ranking factors, and resultant ratings for each of the 12, as well as the total ranking for any or all of the three potentials. NEIC can access the data by Effluent Guideline subcategory, as well as by Standard Industrial Classification Code.
Permit Compliance System (PCS)	A national computerized management information system containing an inventory of NPDES permits, milestone forecasts, inspection events, effluent measurement data, effluent and compliance violations and enforcement actions.	NEIC can acquire limit/measurement data for individual discharges or whole facilities, facilities within a geographic area, sources of a specific industrial classification and the Quarterly Noncompliance Report (QNCR) by Region or State. Information on effluent and compliance schedule violations and enforcement actions/tracking can be obtained.
Pollution Prevention Information Exchange System (PIES)	A national computerized information network providing access to technical, programmatic, and legislative pollution prevention information. Includes a calendar of events, case studies, directory of contacts, an interactive message center, and document ordering capability.	NEIC can use the system to stay abreast of policy and program activities at HQ and the regions as well as industry specific technical information. Case studies of enforcement settlements incorporating pollution prevention projects can be obtained.
Potentially Responsible Parties System	An NEIC-automated system, which links PRPs from SETS, SFFAS, and Techlaw files.	This system is used as an inventory of specific generators or parent corporations identified at and among hazardous waste sites.

AGENCY INTERNAL INFORMATION SYSTEMS
CURRENTLY ACCESSIBLE BY NEIC

System	Description	Application
Records of Decision System (RODS)	A full-text national database of over 2,000 Superfund Records of Decision	NEIC can retrieve a specific ROD by searching onsite name or ID number or can identify all RODS having selected media, contaminants, or remedies.
Resource Conservation and Recovery Information System (RCRIS)	Conversion to RCRIS from HWDMS is currently underway on a per state basis. RCRIS is scheduled to be operating as the official automated source of information on RCRA program activities by January 1992.	NEIC is planning to use the RCRIS National Oversight Database, which is derived from the 10 regional RCRIS databases. Information available will include handler identification, permitting/closure/post-closure, compliance monitoring and enforcement, and corrective action and program management data.
Site Enforcement Tracking System (SETS)	A centralized national database tracking notice letters which have been sent to potentially responsible parties.	NEIC uses this database to supplement currently available responsible party information.
STORET	A national database containing water quality data for some 1,800 unique parameters from over 200,000 collection points including lakes, streams, wells and other waterways. New STORET software provides an interface between STORET and PCS data.	NEIC can access and produce reports of water quality, including groundwater quality, for specific geographical areas, for specific parameters (e.g., organics), and for a specific station.
Superfund Financial Assessment System (SFFAS)	Nationally available computer application designed to calculate the remedial costs a responsible party can theoretically afford to pay for cleanup of a site. Three common financial ratios are used to make this determination: (1) Cash flow to total debt, (2) total debt to equity, and (3) the interest coverage ratio.	NEIC has used the SFFAS to provide financial assessments for potentially responsible parties in response to HQ/Regional requests for several sites including the following: Seymour Recycling (several hundred responsible parties), Re-Solve (more than 200 responsible parties), and MIDCO I and II (approximately 100 responsible parties).

AGENCY INTERNAL INFORMATION SYSTEMS
CURRENTLY ACCESSIBLE BY NEIC

System	Description	Application
TECHLAW Evidence Audit System	Under contract to NEIC, TECHLAW provides document inventories, evidence profiles and generator transaction databases. TECHLAW has produced, for about 30 cases, document inventories containing key word searching capability of all related records contained in Regional office, Headquarters, Department of Justice and/or office files. For sample related activities, including those of the contractor laboratories, TECHLAW produces sample tracking profiles. For hazardous waste sites, TECHLAW has produced document inventories of available records dealing with the generators, volume and type of water, etc.	NEIC can access the document inventories to substantiate the universe of information on which a case is based, to demonstrate the efficacy and utility of an evidence audit system in enforcement case preparation and to provide demonstrative examples of actual applications to establish protocols and implementation procedures.
Toxic Release Inventory System (TRIS)	A national database containing information directly related to the Toxic Chemical Release Inventory Report Form "R." Two types of submissions will be present: Partial (facility and chemical information) and Complete (offsite transfers, emission and releases, waste treatment, waste minimization, activities and uses, and maximum amount stored onsite).	NEIC can generate reports for facilities, geographic areas, and chemical compounds listing facility and chemical information with emissions, releases, activities, etc., for complete submissions.

AGENCY INTERNAL INFORMATION SYSTEMS
NOT CURRENTLY ACCESSIBLE BY NEIC

System	Description	Application
FIFRA/TSCA Tracking System (FTTS)	A PC-based regional system that tracks FIFRA and TSCA inspections, samples, case reviews, enforcement actions, referrals, and State grants. A national database is planned for FY89.	NEIC has limited access through Headquarters. The national database is used to produce facility listings showing inspections, enforcement data, and product lists.

PUBLICLY AVAILABLE EXTERNAL INFORMATION SYSTEMS
CURRENTLY ACCESSIBLE BY NEIC

System	Description	Application
Bibliographic Retrieval Service (BRS)	The BRS system contains more than 100 databases with unique files in the chemical technology and standards and specification areas.	BRS is used mainly as a backup when the DIALOG system is unavailable but is also used to obtain chemical manufacturing and production information for specific compounds.
Chemical Information System (CIS)	The CIS is a collection of scientific and regulatory databases containing numeric, textual and some bibliographic information in the areas of toxicology, environment, regulations, spectroscopy, and chemical and physical properties.	NEIC uses the CIS to locate mass spectral information, environmental fate information, formulation ingredients for commercially available products such as pesticides and waste disposal methods for hazardous substances.
Colorado Alliance of Research Libraries (CARL)	The CARL system includes the catalogs of the member libraries, an index of over 10,000 periodicals, a full text encyclopedia, Choice book reviews, and a bibliography of GPO publications.	CARL is searched by NEIC staff for general reference, to locate books, and to identify articles and documents.
DataTimes	DataTimes provides online access to numerous full-text databases, including newspapers, wire services, and Dow Jones News/Retrieval.	DataTimes is a source of national environmental news. Newspaper databases from all regions are updated daily.
DIALOG Information Services, Inc.	The DIALOG system contains more than 330 databases covering a variety of disciplines: Science, technology, engineering, social sciences, business, and economics. The databases contain more than 120,000,000 records and are regularly updated to provide the most recent information.	NEIC uses the DIALOG databases to obtain: (1) expert witness information, including biographies, publications authored, congressional testimony; (2) up-to-date pollution control technology for hazardous waste, air and water; and (3) business information such as corporate officers, subsidiaries, and line of business.
Dun and Bradstreet	Dun and Bradstreet, a credit-reporting firm, provides business information reports for privately- and publicly-owned companies and government activity reports which list Federal contracts, grants, fines and debarments for specific companies.	NEIC uses the Dun and Bradstreet system to locate corporate information such as business done by the company, company history, financial condition, subsidiaries and corporate officers for privately-held companies.

PUBLICLY AVAILABLE EXTERNAL INFORMATION SYSTEMS
CURRENTLY ACCESSIBLE BY NEIC

System	Description	Application
Groundwater On-Line (GWOL)	The National Groundwater Information Center Database is a bibliographic database containing references to materials on hydrogeology and water well technology with emphasis on reports or projects sponsored by EPA.	NEIC accesses GWOL to locate publications on groundwater topics and to verify or locate groundwater experts.
Justice Retrieval and Inquiry System (JURIS)	The Justice Retrieval and Inquiry System (JURIS) is a legal information system developed by the Department of Justice. It contains the complete text of Federal cases, statutes, and regulations in addition to selected DOJ Appellate Court Briefs, trial briefs, and the full text of the U.S. Attorneys' Manual. The full-text of the NEIC Consent Decree repository has been added to JURIS. JURIS also contains computerized legislative histories of the environmental statutes, model pleadings, and other work products of DOJ attorneys, and EPA General Counsel opinions from 1970 to the present.	JURIS is used to identify relevant caselaw and current statutory and regulatory environmental information.
NEXIS/LEXIS	NEXIS/LEXIS contains the full text of more than 600 business and general news files, including the Washington Post and New York Times. Statutory and case law are provided for computer-aided legal research.	NEIC uses NEXIS/LEXIS to keep informed of the latest Agency and environmental news stories and to track corporate and financial status of U.S. businesses involved in environmental litigation.
National Library of Medicine (NLM)	The National Library of Medicine system contains more than 5,000,000 references to journal articles and books in the health sciences published since 1965.	NEIC uses the NLM system to obtain: (1) toxicity and environmental health effects information for individual chemicals or groups of chemicals, (2) physical and chemical properties of specific compounds, (3) analytical methodology references, and (4) National Cancer Institute carcinogenic bioassay information.

PUBLICLY AVAILABLE EXTERNAL INFORMATION SYSTEMS
CURRENTLY ACCESSIBLE BY NEIC

System	Description	Application
Scientific and Technical Information Network (STN)	The STN system contains databases covering chemistry, science, and engineering that are regularly updated to provide the most recent information. STN has strong coverage of European and Japanese scientific databases.	NEIC uses the STN databases to obtain: (1) chemical structures and synonyms for a chemical compound, (2) analytical methods and techniques, and (3) toxicity of a chemical compound.
VU/TEXT	VU/TEXT contains the full text of 30 daily newspapers, including nationally recognized papers such as the Boston Globe, Chicago Tribune, Detroit Free Press, Philadelphia Inquirer, and regional papers such as the Orlando Sentinel, the Sacramento Bee and the San Jose Mercury News.	NEIC uses VU/TEXT to keep informed of the latest Agency and environmental news stories and to track corporate and financial status of U.S. businesses involved in environmental litigation.
WESTLAW	The WESTLAW system contains legal information, including the full text of cases from the Supreme Court, U.S. Court of Appeals, U.S. District Courts, and State Courts. It contains Shepards' Citations, regulatory information from the Code of Federal Regulations, Federal Register, U.S. Code and the expert witness information from Forensic Services Directory.	NEIC uses WESTLAW to identify precedent cases, to locate all cases decided by a certain judge or all cases represented by a certain attorney and to locate possible expert witnesses.

APPENDIX F

PROJECT PLAN AND SAFETY PLAN EXAMPLES

Example

PROJECT PLAN*
MULTI-MEDIA COMPLIANCE INVESTIGATION
XYZ COMPANY, MIDTOWN, ANYSTATE

<u>INTRODUCTION</u>

The XYZ Company operates a plant at 1234 Anywhere Road in the middle part of Midtown, Anystate [Figure 1]. EPA Region XX requested that NEIC conduct a multi-media compliance investigation of the XYZ plant. The specific objectives of the investigation are to determine compliance with:

- Water pollution control regulations under the Clean Water Act (CWA), including wastewater pretreatment requirements and Spill Prevention and Control Countermeasures (SPCC) regulations

- Hazardous waste management regulations, under the Resource Conservation and Recovery Act (RCRA) and the Anystate Administrative Code (AAC)

- Underground Storage Tank (UST) regulations

- Air pollution control regulations under the Clean Air Act (CAA), Federal Implementation Plan (FIP), and the Federally approved portions of the State Implementation Plan (SIP)

- Toxic Substances Control Act (TSCA) PCB regulations

- Superfund Amendments Reauthorization Act, Title III, Emergency Planning and Community Right-To-Know Act (EPCRA) regulations

Compliance with other applicable environmental regulations may be determined by thew NEIC. Region XX personnel will evaluate compliance with TSCA Sections 5, 8, 12, and 13 during the NEIC inspection, and report their findings separately.

<u>BACKGROUND</u>

XYZ began operating the plant in 1492. Compounds A, B, and C; chemicals D, E, and F; pesticides G and H, and special containers for these materials have been manufactured on site. In 1942, some operations (formerly under the Middle Division) were acquired by a company known as

---

* *Subject to revision*

"Newage, Inc." The remaining XYZ plant currently manufactures water soluble specialty items, and conducts research and development.

The XYZ plant employs a total of about 1,300 people, in a Primary Division, a Secondary Division, a Tertiary Division, and R and D Laboratory. The Primary Division manufactures compounds A, B, and C (240 tons in 1990). Raw materials for the compounds are purchased from an outside source. The Secondary Division makes chemicals and pesticides under numerous brand names (180 tons in 1990), and the Tertiary Division makes special containers for these materials (3 million containers in 1990). Research and development are conducted by R and D Laboratory.

The EPA Region XX Environmental Compliance Division, Midtown District Office (MDO), conducted a multi-media inspection of the XYZ plant during the first quarter of 1991. The MDO inspection report identified concerns with wastewater control, hazardous waste management, documentation, and spill prevention control.

Approximately 1.2 million gallons of wastewater per day are discharged to the Midtown Wastewater Treatment Plant (MWTP) of Midtown, Anystate. There are two direct National Pollutant Discharge Elimination System discharges (001 and 002) to the Midtown River at this facility. Additionally, sewered plant effluent discharge is regulated by the MWTP pretreatment standards, and the Federal effluent limitations and standards for the Compounds, Chemicals, Pesticides and Containers point source category. The R and D Laboratory conducts the Company's effluent analyses.

Violations of the MWTP pretreatment ordinance effluent limitations have occurred for solids, and the toxic standards. MWTP is concerned with data indicating the discharge of solids and toxics J, K, and L from the plant. XYZ also may have modified their pretreatment plant without obtaining a construction permit required by the Anystate Environmental Resources Department (AERD).

XYZ submitted the original RCRA Part A permit application on November 15, 1980. The application listed 19 hazardous waste management units, including 4 container storage areas, 10 storage tanks, and 5 storage surface impoundments. AERD is responsible for monitoring hazardous waste activities.

The facility's June 1990 contingency plan lists 14 above ground and 22 underground tanks on site. The tanks range in size from 2,000 to 50,000 gallons, with the majority between 5,000 and 20,000 gallons. These tanks are located in a tank farm area and near production areas.

The plant emits both volatile organics and particulates. There is no volatile organic constituent emission control equipment. Particulate emissions are controlled by three dust collectors. Five wet scrubbers are used to control fugitive particulate emissions when mixing bags of dry raw materials in reaction vessels. Air emissions are regulated by "Anystate Permits and Air Pollution regulations including AERD Operating Permits. EPA also promulgated a FIP on February 14, 1991.

On August 31, 1983, EPA Region XX conducted a PCB sampling inspection at the plant. XYZ was fined for violations, including cracks in the floor of the PCB storage area, not conducting monthly inspections, no annual document, and not properly marking PCB transformers.

The Toxic Release Inventory (TRI) for this plant lists emissions of A, B, C, and D. The TRI also lists various inorganics, including E, F, G, and H.

<u>INVESTIGATIVE METHODS</u>

Investigation objectives will be addressed by:

- Compilation and review of EPA, AERD, and MWTP database and file information

- Meetings with EPA Region XX personnel to discuss investigation specifics including: objectives, logistics, and potential sampling locations

- An on-site inspection

Meetings with Region XX personnel took place (date). The on-site inspection, scheduled to begin (date), will include:

- Discussing plant operations with facility personnel

- Reviewing and copying, as appropriate, facility documents including operating plans and records

- Visually inspecting plant facilities including processing, material storage, and waste handling facilities

- Sampling and analysis of appropriate waste streams and/or any unknown/unauthorized discharges to assist in compliance determination, as follows:

    (a) MWTP will collect and analyze wastewater samples for organic constituents during the week of (date). All QA/QC will be the responsibility of MWTP.

    (b) NEIC will collect wastewater samples for volatile organic constituent analysis during the on-site inspection. NEIC will conduct the associated analysis.

After completing the on-site inspection, NEIC investigators will brief appropriate EPA Region XX Program and Office of Regional Counsel personnel regarding preliminary findings.

A draft report, including any analytical data, will be written by NEIC personnel and transmitted to EPA Region XX personnel for review and comment. A final report will be completed about two weeks after Region V comments are received. If analytical data are not available by (date), they will be presented in an addendum to the report.

NEIC personnel will be available for any additional support required (negotiations, litigation, etc.) until noncompliance issues are resolved.

DOCUMENT CONTROL PROCEDURES

NEIC document control procedures[*] will be followed during the investigation. TSCA "Notice of Inspection" and "Confidentiality" forms will be completed during the opening conference. Documents and records obtained from the Company will be uniquely numbered and listed on document logs. Photograph logs will also be maintained. A copy of the document and photograph logs, with a Receipt For Samples/Document form, will be offered to the Company prior to completion of the on-site inspection. Any documents declared to be confidential business information pursuant to 40 CFR Part 2 will be so noted on the log and secured appropriately.

SAFETY PROCEDURES

Safety procedures to be followed during the on-site inspection will comply with those described in the attached safety plan [Appendix A], and established NEIC safety procedures. These procedures are contained in EPA 1440 - Occupational Health and Safety Manual (1986 edition), Agency orders, and applicable provisions of the NIOSH/OSHA/USCG/EPA Occupational Safety and Health Guidance Manual for Hazardous Waste Site Activities. The Company's safety policies will also be reviewed and followed.

TENTATIVE SCHEDULE

(date)   Region XX will notify facility of inspection (verbally and in writing)

(date)   Initiate on-site inspection

(date)   Brief Region V regarding preliminary findings

(date)   Draft report to Region V

---

[*]   *NEIC Policies and Procedures Manual, revised August 1991*

APPENDIX

NEIC
SAFETY PLAN
FOR
HAZARDOUS SUBSTANCES RESPONSES AND FIELD INVESTIGATIONS

The OSHA Hazardous Waste Site Worker Standards (29 CFR 1910.120) and EPA protocols require certain safety planning efforts prior to field activities. The following format is aligned with these requirements. Extensive training and certifications are required in addition to this plan.

PROJECT: _________________________________ NEIC Reporting Code: _______

Project Coordinator: _______________________ Date _________________

Branch Chief: _____________________________ Date _________________

On Scene Coordinator or Supervisor:

____

Health and Safety Manager

 Approval: _______________________________ Date _________________

DESCRIPTION OF ACTIVITY

If any of the following information is unavailable, mark "UA"; if covered in project plan, mark "PP".

Site Name: ____
Location and approximate size: _______________________________

Description of the response activity and/or the job tasks to be performed:

____

____

Duration of the Planned Employee Activity: _____________________

____

Proposed Date of Beginning the Investigation: __________________

Site Topography: ____

Site Accessibility by Air and Roads: __________________________

2

HAZARDOUS SUBSTANCES AND HEALTH HAZARDS
INVOLVED OR SUSPECTED AT THE SITE

Fill in any information that is known or suspected

Areas of Concern	Chemical and Physical Properties	Identity of Substance and Precautions
Explosivity:	_________________	_________________
		_________________
		_________________
Radioactivity:	_________________	_________________
		_________________
		_________________
Oxygen Deficiency: (e.g., Confined Spaces)	_________________	_________________
		_________________
		_________________
Toxic Gases:	_________________	_________________
		_________________
		_________________
Skin/Eye Contact Hazards:	_________________	_________________
		_________________
		_________________
Heat Stress:	_________________	_________________
		_________________
		_________________

Pathways from site for hazardous substance dispersion: _________
____
____

WORK PLAN INSTRUCTIONS

A.   Recommended Level of Protection: A ___ B ___ C ___ D ___

     Cartridge Type, if Level C: _________________________________

     Additional Safety Clothing/Equipment: _______________________
____
____
____
____

3

Monitoring Equipment to be Used: ________________________

____
____
____
____

CONTRACTOR PERSONNEL:

Number and Skills ______________________________________

____
____

CONTRACTOR SAFETY CLOTHING/EQUIPMENT REQUIRED: ________

____
____

Have contractors received OSHA required training and certification? (29 CFR 1910.120)

Yes ______   Not Required ______

(If "yes", copy of training certificate(s) must be obtained from contractor)

B.   Field Investigation and Decontamination Procedures:

Decontamination Procedures (contaminated protective clothing, instruments, equipment, etc.): _______________________

____
____
____
____
____
____
____
____
____
____
____

Disposal Procedures (contaminated equipment, supplies, disposal items, washwater, etc.): _________________________

____
____
____
____
____
____
____
____
____
____
____

4

IV.   EMERGENCY CONTACTS

Hospital Phone No.: _______________________________________

Hospital Location: _______________________________________
____
____
____

EMT/Ambulance Phone No.: _______________________________________

Fire Assistance Phone No: _______________________________________

NEIC Health and Safety Manager: Steve Fletcher - 303/236-5111
FTS 776-5111

Radiation Assistance: Wayne Bliss, Director
Office of Radiation Programs
Las Vegas Facility (ORP-LVF)
702/798-2476
FTS 545-2476

APPENDIX G

MULTI-MEDIA INVESTIGATION EQUIPMENT CHECKLIST

Appendix G
MULTI-MEDIA INVESTIGATION EQUIPMENT CHECKLIST

DHL/Federal Express forms

Packing/shipping labels

Packing tape/fiber tape

Custody tape/evidence tape

Coolers with TSCA locks

Cartridges for respirators (14 organic vapor)

Extra vehicle keys

TSCA lock and bar on rear closet doors of van

Accordian folders

Yellow Post-It notes

Box paper clips

Staplers

Boxes of staples

Xerox machine

Related plugs for xerox machine to connect to trailer outlets

Alter 220 extension plug-in to adapt to SCA hookup

Boxes rubber bands

Xerox paper

Writing paper

Box of pens

Box of pencils

Pencil sharpener

Large eraser

Calculators

Lap top/notebook computer

Desk lamps, preferably flat bases

TSCA/PCB forms (CBI green sheets, Confidentiality Notice, Declaration of
    CBI, Notice of Inspection)

Cameras, Polaroid and Nikon 35mm

Boxes film for each (Ecktachrome slides for 35mm)

300-Foot tape

Brunton compass

Two way radios and chargers

Flashlights
Tyvex suits, disposable gloves
Rulers
Desk blotters
Chain-of-custody forms
Ice chests for sample shipping (environmental)
Packing material
NEIC Procedures manual for shipping of samples and TSCA material
Sample receipt forms
Sample tags
Microfilm copier
Microfilm copier film
Tool box
Steel sounding tape
SS weight for tape and SS wire
Boxes Kimwipes
Clipboards
Carpenter's chalk
Sonic sounder
12-Foot tape measure
Garbage bags
Pair NUKs boots
Folding 6-foot ruler
Colored pens/pencils/markers
8-oz. jars
Quart size jars
Plastic Ziplocks
Glass thieves
Plastic/metal scoops
Shovel
pH paper/meter
HNu meter
Bacon bombs
Sampling gear
Media-specific sampling gear

APPENDIX H

SUGGESTED RECORDS/DOCUMENTS REQUEST

Appendix H
SUGGESTED RECORDS/DOCUMENTS REQUEST

<u>GENERAL PROCEDURE</u>

The records evaluation generally will proceed in two stages. First, various records to be reviewed will be identified. Generally, these records will date back three years from the present, but some of the records will be for specific time periods. Second, according to a schedule to be developed onsite, the records will be reviewed and copies requested, as needed. Alternately, document copies will be requested for later review after the investigation.

<u>GENERAL</u>

1. Facility map and plot plan
2. Organizational chart
3. Description of facility and operations

<u>CLEAN AIR ACT (CAA)</u>

1. Plot plan of the facility showing location and identification of all major process area and stacks

2. Brief descriptions for all process areas to include:

   (a) simplified process flow diagrams
   (b) pollution control equipment

3. Permits and/or variances for air emission sources and related correspondence

4. Consent Decrees/Orders/Agreements still in effect

5. Sulfur in fuel records for boiler/space heater fuel

6. Stack tests (most recent) and stack and ambient monitoring data (last 2 quarters)

7. Performance specification tests for continuous emission monitors

8. 1989 state emissions inventory report

9. Procedures/manuals for the operation and inspection of pollution control equipment

10. Required notices for asbestos demolition/renovation projects in progress or completed within the last three years

11.   Hours of operation and process weight rates for the automated multi-base propellant (AMB) manufacturing facility (last 2 years)

12.   Annual volatile organic compounds (VOC) emissions from the AMB facility including associated VOC storage tanks (last 2 years). Describe the method(s) used to determine emissions. If estimates have been made show calculations and assumptions.

13.   Annual AMB facility inspections reports (last 2 years)

14.   Annual AMB facility VOC emissions control reports (last 5 years)

15.   Non-compliance AMB reports (last 2 years)

16.   List air pollution sources, not covered in items 1 - 8, such as combustions units larger than 2.5 million BTUs/hr heat inputs and incinerators (other than the RCRA incinerators)

RESOURCE CONSERVATION AND RECOVERY ACT (RCRA)

1.   List of description of all hazardous waste storage areas, including above and below ground tanks, temporary tanks, drum storage areas, pits, ponds, lagoons, waste piles, etc., that have been operated at any time since November 1980.

2.   RCRA Part A Permit Application

3.   Manifests for all offsite shipments of hazardous waste, including notifications and certifications for Land Disposal Restricted (LDR) hazardous waste.

4.   Determinations, data, documents, etc., supporting the Base's decision that wastes are hazardous, non-hazardous or LDR hazardous wastes for all solid wastes, as defined under RCRA. Also provide information used in the determination of the EPA hazardous waste codes applied to all hazardous wastes.

5.   Notices to the owner or operator of the disposal facility receiving waste subject to land disposal restrictions

6.   Schedule and inspection logs for inspection of safety and emergency equipment, security devices, monitoring equipment, and operating and structural equipment

7.   Satellite accumulation area inspection records

8. Employee training records for hazardous waste handlers, including job title and description, name of each employee and documentation of the type and amount of training each has received.

9. Current Contingency Plan

10. Current Closure Plan

11. Copy of the Waste Analysis Plan (WAP) currently in use and effective date of the plan. If the current WAP was not effective on January 1, 1986 provide copies of all WAPs and revisions.

12. Narrative of procedures used to store hazardous wastes prior to shipment offsite for treatment, recycle, reclamation and/or disposal. Include a list of all storage and satellite storage areas and the quantity of waste stored at each area.

13. Summary reports and documentation of all incidents that required implementation of the contingency plan for the past 3 years

14. The Generator Biennial and Exception Reports

15. Reports and analytical results of any groundwater quality and groundwater contamination surveys

16. Closure plan for units undergoing closure

17. Inspection schedule(s) for all hazardous waste management units, such as storage areas and tank systems, and all inspection logs, remediations document, etc, for the last three years.

18. Description of the Base's hazardous waste minimization plan

19. Copy of the annual report for the last three years

20. Copy of notification to EPA of hazardous waste activity to secure generator ID number

21. Copy of Clean Closure Demonstration, if any.

22. List of all identified or suspected Solid Waste Management Units on the Base's property

23. List of all locations where hazardous wastes are generated including types, quantity, and EPA hazardous waste codes of wastes

24. Notification of any releases to the environment and follow-up reports

25. Notifications for underground storage tanks

26. Copy of written tank integrity assessment certified by a professional engineer

27. Agreements with local emergency response authorities or documentation of refusal by the emergency response authorities to enter into such agreements.

28. Design specifications for any underground storage tanks installed after May 1985

29. Characteristics of materials removed from facility septic systems, including analytical results

30. Wastes burned in the RCRA incinerators, including waste analysis

31. Instruments used to monitor incinerator combustions and emissions

32. Incinerator inspection records (last year)

TOXIC SUBSTANCES CONTROL ACT - PCB WASTE MANAGEMENT

1. Copies of the "Annual Document" required by 40 CFR 761.180(a) for the last 3 years

2. Records of monthly inspections of storage areas subject to 40 CFR 761.65

3. The SPCC plan prepared for storage areas subject to 40 CFR 761.65

4. All spill reports

5. Records of inspection and maintenance for PCB transformers for the last 3 years

6. Transformer inventory and PCB analyses

7. Reports or other documentation identifying the extent of any PCB spills and any remediation plans

8. Certifications of Destruction for PCB Transformer disposal

9. Manifests for PCB items shipped off-site

CLEAN WATER ACT (CWA)

1. All NPDES permit(s) applications

2. NPDES permit effective for last 3 years

3. Discharge monitoring reports (DMRs) for last 3 years

4. Any correspondence regarding exceedences of discharge limitations

5. Spill Prevention Control and Countermeasure (SPCC) plan and Prevention Preparedness and Contingency (PPC) plan

6. Description or lies of all sewer system monitoring stations and analyses conducted on samples collected (include monitoring frequency).

7. Written calibration procedures for flow measuring and recording equipment; includes industrial sewers, storm sewers, sanitary sewers or any other sewers on the Plant's property.

8. Description of waste water treatment plant, sewer system and storm water by-pass system

9. Consent decrees/agreements still in effect

COMPREHENSIVE ENVIRONMENTAL RESPONSE COMPENSATION AND LIABILITY ACT (CERCLA)

1. Specific CERCLA questions will be provided during the inspection.

EMERGENCY PLANNING COMMUNITY RIGHT-TO-KNOW ACT (EPCRA)

1. Notification to the State Emergency Response Commission

2. Designated facility emergency coordinator

3. Written follow-up emergency release notifications

4. Material Safety Data Sheet report to the Commission and fire department

5. Tier I/Tier II submittal to the Commission and fire department

6. EPA Form R submittals for the past three years

7. Documentation supporting the Form R submittals for the past three years

FEDERAL INSECTICIDE, FUNGICIDE, AND RODENTICIDE ACT (FIFRA)

1.   Restricted pesticide use and application records

2.   Pesticide handlers training and certification records

3.   Pesticide inventory and storage area inspection records

SAFE DRINKING WATER ACT

1.   Description of water supply system including supply, storage capacity, distribution, and monitoring system

APPENDIX I

MULTI-MEDIA CHECKLISTS

Appendix I
MULTI-MEDIA CHECKLISTS

TITLE	DATE	SOURCE
Multi-Media		
Facility Multi-Media Survey		Region I
Inspector's Multi-Media Checklist		Region II
RCRA		
Pre-Inspection Worksheet	June 1989	RCRA Inspection Manual [*]
Land Ban Checklist Questions	June 1989	RCRA Inspection Manual
General Site Inspections Information Form	June 1989	RCRA Inspection Manual
General Facility Checklist	June 1989	RCRA Inspection Manual
Land Disposal Restrictions List	June 1989	RCRA Inspection Manual
RCRA Hazardous Waste Tank Inspec.	June 1989	RCRA Inspection Manual
Transporter's Checklist	June 1989	RCRA Inspection Manual
Containers Checklist	June 1989	RCRA Inspection Manual
Surface Impoundments List	June 1989	RCRA Inspection Manual
Waste Piles Checklist	June 1989	RCRA Inspection Manual
Land Treatment Checklist	June 1989	RCRA Inspection Manual
Landfills Checklist	June 1989	RCRA Inspection Manual
Incinerators Checklist	June 1989	RCRA Inspection Manual
Thermal Treat. List(part 264)	June 1989	RCRA Inspection Manual
Ground-Water Monitoring Checklist	June 1989	RCRA Inspection Manual
Waste Information Worksheet	June 1989	RCRA Inspection Manual
Comprehensive GW Monitoring Eval.	March 1988	RCRA GW Monitoring Systems Manual
Comparison of Permit & Oper. Cond.	April 1989	RCRA Incinerator Inspection Manual
Visual Assess. & Audit Activities	April 1989	RCRA Incinerator Inspection Manual
List-Inspec. New RCRA Incinerators	April 1989	RCRA Incinerator Inspection Manual
Landfill and Dump Site Analysis	April 1989	RCRA Incinerator Inspection Manual
Chemical Facility analysis	June 1988	RCRA Incinerator Inspection Manual
RCRA Land Disposal Rest. Gen. List	Feb. 1989	RCRA Incinerator Inspection Manual
Transporter Checklist	Feb. 1989	RCRA Incinerator Inspection Manual
RCRA Land Restrictions-T,S & D Req.	Feb. 1989	RCRA Incinerator Inspection Manual
Solvent Identification Checklist	Feb. 1989	RCRA LDR Inspection Manual
Tank Systems-Inspection Checklist	Sept. 1988	RCRA Tank Inspection Manual
Tank System-Small Quantity Gen.	Sept. 1988	RCRA Tank Inspection Manual

---

* These manuals are available from Government Institutes, Inc. Please see back page. (03/92)

Appendix I (cont.)
MULTI-MEDIA CHECKLISTS

TITLE	DATE	SOURCE
Tank-Document of General Inspection Requirements	Sept. 1988	RCRA Tank Inspection Manual
Tank System-Existing Tank System	Sept. 1988	RCRA Tank Inspection Manual
Tank System-New Tank System	Sept. 1988	RCRA Tank Inspection Manual
Tank Systems and Ignitable Waste	Sept. 1988	RCRA Tank Inspection Manual
Tank System-Release Response	Sept. 1988	RCRA Tank Inspection Manual
Tank System-Visual Tank Inspec	Sept. 1988	RCRA Tank Inspection Manual
Tank-Closure, Post-Closure Call	Sept. 1988	RCRA Tank Inspection Manual
Landfill & Dump Site Analysis	June 1988	RCRA Tank Inspection Manual
Chemical Facility Analysis	June 1988	RCRA Tech. Case Devel Guidance
RCRA LDR Inspections	Sept. 1990	Prepared by Region V
Health & Safety Inspection Form	Feb. 1991	OWPE
TSCA		
TSCA Screening Inspection Checklist		Region VI
Established Inspections Narrative Report	Jan. 1989	Pesticides Inspection* Manual
CWA		
Record, Report & Schedule List	1985	CWA* Compliance/Enforcement Guidance
Sample Evaluations List Sect. 309	1985	CWA Compliance/Enforcement Guidance
NPDES		
NPDES Compliance Inspection Report		NPDES Compliance Monitoring - Overview
Mobile Bioassay Equipment List		NPDES Compliance Monitoring - BioMonitoring
Records, Reports & Schedule List	June 1984	NPDES Compliance* Inspection Manual
Facility Site Review Checklist	June 1984	NPDES Compliance Inspections Manual
Permittee Sampling Inspec.	June 1984	NPDES Compliance Inspection Manual
Flow Measurement Inspection List	June 1984	NPDES Compliance Inspection Manual

---

* These manuals are available from Government Institutes, Inc. Please see back page. (03/92)

Appendix I (cont.)
MULTI-MEDIA CHECKLISTS

TITLE	DATE	SOURCE
NPDES Biomonitoring Evaluation Form	June 1984	NPDES Compliance Inspection Manual
Laboratory Quality Assurance List	June 1984	NPDES Compliance Inspection Manual
UIC		
Inspections Checklist (UIC)	Feb. 1988	UIC Inspection Manual for U.S. EPA
Pressure Gauge Inspection List	Feb. 1988	UIC Inspection Manual for U.S. EPA
Flow Measurement Inspec. List	Feb. 1988	UIC Inspection Manual for U.S. EPA
Inspections Checklist (UIC)	Feb. 1988	UIC Inspection Manual for U.S. EPA
Pressure Gauge Inspection List	Feb. 1988	UIC Inspection Manual for U.S. EPA
Flow Measurement Inspection List	Feb. 1988	UIC Inspection Manual for U.S. EPA
Air		
Electric Arc Furnaces (I) List	May 1977	Steel Producing Electric Arc Furnaces
Opacity Observations (II) List	May 1977	Steel Producing Electric Arc Furnaces
Performance Test Observation (III)	May 1977	Steel Producing Electric Arc Furnaces
Operation of Electric Arc Furnace (IV)	May 1977	Steel Producing Electric Arc Furnaces
Fume Collection System (V) Checklist	May 1977	Steel Producing Electric Arc Furnaces
Fabric Filter Collectors (VI) Checklist	May 1977	Steel Producing Electric Arc Furnaces
Scrubbers (VI) Checklist	May 1977	Steel Producing Electric Arc Furnaces
Electrostatic Precipitators (VIII) List	May 1977	Steel Producing Electric Arc Furnaces
Tank Inspection Checklist	April 1977	Volatile Hydrocarbon Storage Tanks
Sewage Sludge Incinerators -During Performance Test	Feb. 1975	Sewage Sludge Incinerators
Sewage Sludge Incinerators-After Performance Test	Feb. 1975	Sewage Sludge Incinerators
Municipal Incinerators - During Performance Test	Feb. 1975	Municipal Incinerators
Municipal Incinerators - After Performance Test	Feb. 1975	Municipal Incinerators

Appendix I (cont.)
MULTI-MEDIA CHECKLISTS

TITLE	DATE	SOURCE
Secondary Brass & Bronze Smelters - During Performance Test	Jan. 1977	Secondary Brass & Bronze Ingot Production Plants
Secondary Lead Smelters - During Performance Test	Jan. 1977	Secondary Lead Smelters
Basic Oxygen Process Furnace - During Performance Test	Jan. 1977	Basic Oxygen Process Furnaces
Performance Test of Portland Cement Plants	Sept. 1975	Portland Cement Plants
Periodic Check of Portland Cement Plant	Sept. 1975	Portland Cement Plants
Steam-Electric Generators-During Performance Test	Feb. 1975	Fossil-Fuel Fired Steam Generators
Steam-Electric Generation -After Performance Test	Feb. 1975	Fossil-Fuel Fired Steam Generators
Municipal Incinerators Checklist	June 1973	Combustion & Incineration Sources

APPENDIX J

INVESTIGATION AUTHORITY/SUMMARY OF
MAJOR ENVIRONMENTAL ACTS

Table 1

INVESTIGATION AUTHORITY UNDER THE MAJOR ENVIRONMENTAL ACTS

CAA - § 114(a)(2)

". . .the Administrator or his authorized representative, upon presentation of his credentials - shall have a right of entry to, upon, or through any premises of such person or in which any records required to be maintained. . .are located, and may at reasonable times have access to and copy any records, inspect any monitoring equipment and method. . .and sample any emissions. . . ."

CWA - § 308(a)(4)(B)

". . .the Administrator or his authorized representative. . .upon presentation of his credentials - (i) shall have a right of entry to, upon, or through any premises in which an effluent source is located or in which any records required to be maintained. . .are located, and (ii) may at reasonable times have access to and copy any records, inspect any monitoring equipment or method. . . any sample any effluents which the owner or operator of such source is required to sample. . . ."

RCRA - § 3007(a)

". . .any such person who generates, stores, treats, transports, disposes of or otherwise handles or has handled hazardous wastes shall upon request of any. . .employee or representative of the Environmental Protection Agency. . .furnish information relating to such wastes and permit such person at all reasonable times to have access to, and to copy all records relating to such wastes."

". . .such employees or representatives are authorized. . .to enter at reasonable times any establishment or other place where hazardous wastes are or have been generated, stored, treated or disposed of or transported from; to inspect and obtain samples from any person of any such wastes and samples of any containers or labeling for such wastes."

- § 9005(a)(1)

". . .representatives are authorized. . .to enter. . .inspect and obtain samples. . .

TSCA - § 11(a)(b)

". . .any duly designated representative of the Administrator, may inspect any establishment. . .in which chemical substances or mixtures are manufactured, processed, stored or held before or after their distribution in commerce and any conveyance being used to transport chemical substances, mixtures or such articles in connection with distribution in commerce. Such an inspection may only be made upon the presentation of appropriate credentials and of a written notice to the owner, co-operator or agent in charge of the premises or conveyance to be inspected."

FIFRA - § 8 and 9

". . .any person who sells or offers for sale, delivers or offers for delivery any pesticide. . .shall, upon request of any officer or employee of the Environmental Protection Agency. . .furnish or permit such person at all reasonable times to have access to, and to copy: (1) all records showing the delivery, movement or holding of such pesticide or device, including the quantity, the date of shipment and receipt, and the name of the consignor and consignee. . . ."

". . .officers or employees duly designated by the Administrator are authorized to enter at reasonable times, any establishment or other place where pesticides or devices are held for distribution or sale for the purpose of inspecting and obtaining samples of any pesticides or devices, packaged, labeled and released for shipment and samples of any containers or labeling for such pesticides or devices."

"Before undertaking such inspection, the officers or employees must present to the owner, operator or agent in charge of the establishment. . . appropriate credentials and a written statement as to the reason for the inspection, including a statement as to whether a violation of the law is suspected."

". . .employees duly designated by the Administrator are empowered to obtain and to execute warrants authorizing entry. . .inspection and reproduction of all records. . .and the seizure of any pesticide or device which is in violation of this Act."

SDWA - §1445

". . .the Administrator, or representatives of the Administrator. . .upon presenting appropriate credentials and a written notice to any. . .person subject to. . .any requirement. . .is authorized to enter any establishment, facility or other property. . .in order to determine. . .compliance with this title, including for this purpose, inspection, at reasonable times, of records, files, papers, processes, controls and facilities or in order to test any feature of a public water system, including its raw water source."

CERCLA (Superfund) - § 104(e)

"Any officer, employee or representative of the President. . .is authorized to. . .

require any person. . .to furnish. . .information or documents relating to. . .identification, nature and quantity of material. . .generated, treated, stored, or disposed. . .or transported. . .nature or extent of a release. . .ability of a person to pay. . ."

". . .access. . .to inspect and copy all documents or records. . ."

". . .to enter. . .place or property where any hazardous substance or pollutant or contaminant may be or has been generated, stored, treated, disposed of, or transported from. . .needed to determine the need for response. . ."

". . .to inspect and obtain samples. . ."

Table 2

SUMMARY OF FEDERAL ENVIRONMENTAL ACTS REGARDING RIGHT OF ENTRY, INSPECTIONS, SAMPLING, TESTING, ETC.

Act/Section	Designated Representative	Presentation of Credentials	Notice of Inspection	Sampling Permitted	Inspection of Records	Sample Splits	Receipt for Agency's Samples	Return of Analytical Results
Clean Water Act - § 308(a)	Yes, auth. by Administrator	Required	Not required	Yes (effluents which the owner is required to sample)	Yes	Not required	Not required	Not required
FIFRA - § 8(b) (Books and Records)	Yes, designated by Administrator	Required	Written notice required with reason and suspected violation note	Access and copy records	Yes	N/A	N/A	N/A
FIFRA - § 9(a) (Inspections of Establishments)	Yes, designated by Administrator	Required	Written notice required with reasons for for inspection	Yes	See § 8	Required, if requested	Required	Required, promptly
Clean Air Act - § 114(a)	Yes, auth. by Administrator	Required	Not required except notify State for SIP sources	Yes	Yes	Not required	Not required	Not required
RCRA - § 3007(a) § 9005(a)	Yes, designated by Administrator	Not required	Not required	Yes	Yes	Required, if requested	Required	Required, promptly
SDWA - §1445(b)	Yes, designated by Administrator	Required	Written notice required, must also notify State with reasons for entry if State has primary enforcement responsibility	Yes	Yes	Not required	Not required	Not required
TSCA - § 11(a, b)	Yes, designated by Administrator	Required	Written notice required	(The Act does not mention samples or sampling in this section. It does state an inspection shall extend to all things within the premise of conveyance.)	Yes	N/A		N/A
CERCLA - §104(e)	Yes, designated by President	Not required	Upon reasonable notice for information	Yes	Yes	Required, if requested	Required	Required, promptly

APPENDIX K

EVIDENTIARY PROCEDURES FOR PHOTOGRAPHS/MICROFILM

Appendix K
PHOTOGRAPHS

When movies, slides or photographs are taken which visually show the effluent or emission source and/or any monitoring locations, they are numbered to correspond to logbook entries. The name of the photographer, date, time, site location and site description are entered sequentially in the logbook as photos are taken. A series entry may be used for rapid sequence photographs. The photographer is not required to record the aperture settings and shutter speeds for photographs taken within the normal automatic exposure range. Special lenses, films, filters, or other image enhancement techniques must be noted in the logbook. Chain-of-custody procedures depend upon the subject matter, type of film, and the processing it requires. Film used for aerial photography, confidential information or criminal investigations require chain-of-custody procedures. Adequate logbook notations and receipts may be used to account for routine film processing. Once developed, the slides or photographic prints shall be serially numbered corresponding to the logbook descriptions and labeled.

MICROFILM

Microfilm is often used to copy documents that are or may later become TSCA Confidential Business Information (CBI). This microfilm must be handled in accordance with the TSCA CBI procedures (see Appendix I for additional information and forms). Table C-1 is the NEIC procedure for processing microfilm containing TSCA CBI documents.

Table K-1

NEIC PROCEDURE FOR MICROFILM
PROCESSING OF TSCA CBI DOCUMENTS

1. Kodak Infocapture AHU 1454 microfilm shall be used for filming all TSCA CBI documents.

2. Obtain packaging materials and instructions from the NEIC Document Control Officer or Assistant, including:

    - Preprinted shipping labels
    - Chain-of-custody records
    - Custody seals
    - Double envelopes
    - Green TSCA cover sheets
    - TSCA loan receipt

3. Prepare each roll of microfilm for shipment to the processor.

    - Enclose the film in double-wrapped packages
    - Place a green TSCA cover sheet in the inner package
    - Place a TSCA loan receipt in the inner package
    - Complete a Chain-of-Custody Record, place the white copy in the inner package and keep the pink copy for the field files
    - Seal inner package with a custody seal and sign and date it
    - Mark the inner package:

    "TO BE OPENED BY ADDRESSEE ONLY
    TSCA CONFIDENTIAL BUSINESS INFORMATION"

4. Ship the film via Federal Express to the Springfield, Virginia Federal Express office and instruct that it is to be held for pickup. USE SIGNATURE SECURITY SERVICE ONLY.

    This practice requires the courier to sign, the station personnel to sign, and the delivery courier to sign.

    Instruct the Springfield Federal Express office to hold the shipment for pickup and to notify:

    Mr. Vern Webb
    U.S. EPA/EPIC
    Vint Hill Farms Station
    Warrenton, Virginia 22186
    (730) 557-3110

5. Telephone Mr. Webb and inform him of the date shipped, the number of rolls of film, the air bill number, and your phone number.

6. Telephone the NEIC Document Control Officer or Assistant and inform them.

7. Telephone Mr. Webb the following day and verify film quality to determine if repeat microfilming is necessary.

8. The pink copy of the Federal Express form, with the shipment cost and project number indicated, must be turned in to the Assistant Director, Planning and Management. If you are in the field for an extended period of time (3 weeks or more), the pink copies must be mailed to NEIC.

APPENDIX L

KEY PRINCIPLES AND TECHNIQUES FOR INTERVIEWING

KEY PRINCIPLES AND TECHNIQUES FOR INTERVIEWING

The list of principles and techniques presented below is intended to highlight methods which can be used by auditors to conduct effective interviews.

Planning the Interview

- Iron out logistics
- Define the desired outcome(s)
- Organize thoughts and establish a general sequence for questioning

Opening the Interview

- Introduce yourself and the purpose of the interview
- Ensure appropriateness of time
- Explain how information will be used

Conducting the Interview

- Request a brief overview of the interviewee's responsibilities with respect to the audit topic(s)
- Ask open-ended questions (e.g., "what" or "how"), not obvious yes/no questions (e.g., "do you", etc.)
- Follow-up on issues which are unclear
- Avoid making assumptions
- Avoid leading questions
- Provide feedback to interviewee questions, as appropriate, to ensure a level of responsiveness to the interviewee.
- Tolerate silences in order to allow the interviewee to formulate thoughts and responses

Closing the Interview

- Do not exceed the agreed-upon time limit without getting concurrence for an extension
- End on a positive note
- Summarize your understanding of key points discussed to ensure accuracy

Documenting Interview Results

- Establish the context of the interview (time, name of interviewee, protocol step)
- Take notes of key points during the interview (do not attempt a verbatum transcript)
- Summarize the outcome and overall conclusions at the end of the audit

Interpersonal Considerations

- Use appropriate voice tone and inflection
- Do not jump to conclusions

Interview Setting

- Make sure the interviewee feels that there is sufficient privacy
- When appropriate, conduct the discussion in the interviewee's work area
- Try to keep it "one-on-one"
- Minimize distractions

Non-Verbal Communication

- Shake hands
- Maintain eye contact
- Keep the right distance
- Mirror the interviewee

APPENDIX M

SAMPLING GUIDELINES

Appendix M
SAMPLING GUIDELINES

The value of samples as evidence to document/support a violation is contingent upon many factors including: (1) the method by which samples are collected; (2) the selection of sample containers, preserving samples after they are collected, and ensuring that proper holding times are adhered to between sample collection and analysis; (3) the accuracy or validity of field measurements that are taken in conjunction with that sampling; (4) the adequate decontamination of field sampling equipment; (5) the degree to which appropriate notes or other documentation pertaining to sampling operations are logged in a notebook; and (6) the labeling of samples and employing a suitable chain-of-custody system. Most of these topics will be discussed in a relatively brief fashion in this section of the document. For additional details, the reader should refer to any of the SOPs currently being used by the Regions or NEIC. Other useful documents are program specific sampling/analytical protocols, such as 40 CFR Part 136 (NPDES), SW 846 (RCRA), and the Technical Enforcement Guidance Document (RCRA).

SAMPLE COLLECTION

Samples can generally be divided into two separate and distinct categories: (1) environmental samples and (2) source or waste samples. The collection of both will probably be necessary during most multi-media inspections. Environmental samples can include surface/runoff water, groundwater, sediment, surface wipes, soil, etc. Source or waste samples can include discharges from permitted outfalls, PCB oil from electrical equipment, RCRA regulated "hazardous waste", treatment residues, leachate, etc. In the case of any toxic/hazardous materials, the inspectors should make every effort to have the facility collect the sample for them. Sampling of either category can be accomplished by collecting grab samples, composite samples, or both. The type of sample ultimately obtained will be predicated on satisfying a legal requirement such as a permit which specifies a type of sample, laboratory requirements, or ensuring that representativeness is achieved. A time-based composite sample will usually require the use of an automatic sampler set to collect a series of discrete samples, over the time period of interest. A spatially-based composite sample (actually a series of grab samples blended together) and grab samples are collected by more conventional means and during a much shorter time. Some type of dipper, scoop, auger, pump, corers, etc. can be used to collect a grab sample.

Perhaps the two most important points the investigators should keep in mind whenever sampling, are identifying precisely where the sample will be collected and selecting the appropriate equipment to collect the sample. From a collection standpoint, a sample must often be obtained such that it is representative of the entire media. If the media is well mixed and homogenous, a single sample will probably be adequate to ensure representativeness. If it is not well mixed, the investigator will have to collect several samples at different locations, and composite them on an equally weighted or some other basis or have each of the discrete samples analyzed. The total number of samples required will largely depend upon the area/volume of the material and the degree of nonhomogeneity. Normally in this case, the investigator will have to use a statistically random process to determine where the samples should be collected. Laying out some type of imaginary grid to encompass the media to be sampled and randomly selecting specific elements for sampling is one common method of ensuring that bias is not introduced and that no other important statistic is compromised. In other cases, the inspector must rely on a judgmental sampling approach, particularly in situations where a worst case result is desired. In that situation, the inspector should look for signs of discoloration, wetness, waste plumes or residue, dead vegetation, odor, or some other physical attribute or apply knowledge of the situation (i.e., judgment) in an attempt to identify exactly where constituent concentrations are likely to be highest. Other items that must be considered when selecting sampling sites are safety, convenience, and accessibility. The investigators should not collect samples until they have adequate knowledge of the site, through touring/observing, interviewing, etc. to make prudent decisions regarding selection of sampling sites.

The second important point the investigator should remember is to use the proper equipment to physically collect the sample. The equipment should either be unused or decontaminated to an extent that it cannot impart any contamination to the sample itself. Moreover, it is important to select sampling equipment made of the proper material. Wherever possible, the material should be inert (i.e., teflon or stainless steel) and not contain, as its principal constituents, any of the same constituents for which the samples will be analyzed. Lastly, the element of safety should not be overlooked. One key example would be where the investigators need to sample a potentially hazardous waste from a 55-gallon drum. Opening the drum must be carefully performed prior to sampling.

SAMPLE CONTAINERS, PRESERVATION METHODS, HOLDING TIMES, BLANKS

In order for the samples to be properly analyzed in the laboratory, the field investigators must follow certain accepted procedures relative to the containers they use, the preservation of the samples at time of collection, the holding time limits which dictate the quickness in which the samples must be transported to the laboratory, and the use of field blanks for QA purposes. Each of these procedures can vary from matrix to matrix, parameter to parameter, and in some cases, from program to program. All of this information pertaining to sample collection/handling is summarized in the following two tables. One table corresponds to liquid samples and the other to nonliquid samples.

FIELD MEASUREMENTS

The acquisition of field data is customarily required whenever sampling is performed. A number of *in situ* monitoring devices/meters are used for this purpose, each of which have certain applications and limitations. These instruments are designed to withstand some rough handling without affecting their stability or reliability. Parameters normally measured in the field include but, are not limited to, dissolved oxygen, pH, temperature, and chlorine residual.

SAMPLE COLLECTION/HANDLING REQUIREMENTS - LIQUID SAMPLES

Parameter	Sample Type	Container Size/Type	Minimum Volume	Preservation	Holding Times	Blanks[4]
Bacteriological						
Coliform, Fecal & T	Grab	250 ml P,G sterilized	3/4 full	Cool, 4 C - 0.2ml, 10% $Na_2S_2O_3$ added in lab	6-hr.	Field (A)
Inorganic						
BOD		1000 ml P,G	400 ml	Cool, 4°C	48-hr.	None
COD		1000 ml P,G	200 ml	Cool, 4°C-H_2SO_4 to pH<2	28 day	Acid
TSS		1000 ml P,G	100 ml	Cool, 4°C	7 day	None
TKN		1000 ml P,G	200 ml	Cool, 4°C-H_2SO_4 to pH<2	28 day	Acid
Ammonia				Cool, 4°CH_2SO_4 to pH<2	28 day	Acid
Nitrite		1000 ml P,G	400 ml	Cool, 4°C	48 hr	None
Nitrate				Cool, 4°C	48 hr	None
NO_2 NO_3		1000 ml P,G	200 ml	Cool, 4°C - H_2SO_4 to pH<2	28 day	Acid
Ortho Phosphorus		1000 ml P,G	200 ml	Cool, 4°C	48 hr	None
Total Phosphorus		1000 ml P,G	200 ml	Cool, 4°C - H_2SO_4 to pH<2	28 day	Acid
Oil & Grease	Grab	1000 ml G only	1000 ml	Cool, 4°C - H_2SO_4 to pH<2	28 day	Acid
Sulfide	Grab	300 ml G		Cool, 4°C - Zinc Acetate; NaOH to pH>9	7 day	Field
Cyanide T & Amen	Grab	1000 ml P	500 ml	Cool, 4°C - NaOH to pH>12[3]	14 day	NaOH
Hex Chrome	Grab	1000 ml P	300 ml	Cool, 4°C	24 hr	Field
Mercury		1000 ml P	500 ml	HNO_3 to pH<2	28 day	Acid
Other Heavy Metals		1000 ml P	500 ml	HNO_3 to pH<2	6 mo	Acid
Organic						
Volatiles	Grab	3-40 ml G vials[1]	40 ml	Cool, 4°C-HCl to pH<2,Dechlor w/25 mg/40ml Ascorbic Acid	7 day-Unpres. 14 day-Pres.	Field(3)
Extract, P.P.s		2-1/2gl G(ambr)[1]	1000 ml	Cool,4°C-Dechlor w/1.0 ml $Na_2S_2O_3$	7 day[2]	
Extract Pest/Herb		2-1/2gl G(ambr)[1]	1000 ml	Cool,4°C-Dechlor w/1.0 ml $Na_2S_2O_3$	7 day[2]	Field(1)
Phenols		1-qt. G[1]	1000 ml	Cool 4°C-H_2SO_4 to pH<2	28 day	Acid
PCB		1-40ml G-Vial[4]	2 ml[4]	Cool, 4°C	7 day[2]	None
TOC		1000 ml P	200 ml	Cool, 4°C-H_2SO_4 to pH<2	28 day	Acid
TCLP						
Volatiles	Grab	3-40 ml G vials[1]	40 ml	Cool, 4°C	14 day[2]	None
Semi-Volatiles		Qt. Mason Jar[1]	500 ml	Cool, 4°C	14 day[2]	None
Mercury		1000 ml P	500 ml	Cool, 4°C	28 day[2]	None
Other Metals		1000 ml P	500 ml	Cool, 4°C	180 day[2]	None

[1] Teflon Lid Liners

[2] To Extraction

[3] Check for Cl_2 Residual and Sulfides (lead acetate paper). If Cl_2 is present, add ascorbic acid. If sulfide is present, remove with cadmium nitrate powder, filter and pH - 12 with NaOH.

[4] For oil samples only. Water samples require 1000 ml and can be collected in a qt. mason jar with teflon lid liner.

Sample Collection/Handling Requirements - Non-Liquid Samples

Parameter	Sample Type	Container Size/Type	Minimum Volume	Preservation	Holding Times	Blanks
Inorganic						
COD		8 oz. glass jar[1]	5 gr.	Cool, 4° C	28 days	None
TKN		8 oz. glass jar[1]	5 gr.	Cool, 4° C	28 days	None
Total Phosphorus		8 oz. glass jar[1]	5 gr.	Cool, 4° C	28 days	None
Cyanide	Grab	8 oz. glass jar[1]	5 gr.	Cool, 4° C	14 days	None
Mercury		8 oz. glass jar[1]	0.2 gr.	Cool, 4° C	28 days	None
Other Heavy Metals		8 oz. glass jar[1]	0.2 gr.	Cool, 4° C	6 mos.	None
Organic						
Volatiles	Grab	2-40 ml G vials[1]	40 ml[3]	Cool, 4° C	7 days	Field (3)
Extract P.P.s		8 oz. glass jar[1]	100 gr.	Cool, 4° C	10 days[2]	None
Extract Pest/Herb		8 oz. glass jar[1]	100 gr.	Cool, 4° C	10 days[2]	None
Phenols		8 oz. glass jar[1]	100 gr.	Cool, 4° C	28 days	None
PCB		8 oz. glass jar[1]	10 gr.	Cool, 4° C	10 days[2]	None
TCLP						
Volatiles	Grab	2-40 ml G Vials[1]	40 ml[3]	Cool, 4° C	14 days[2]	None
Semi-Volatiles		8 oz. glass jar[1]	200 gr.-wet	Cool, 4° C	14 days[2]	None
Mercury		8 oz. glass jar[1]	)200 gr.-wet	Cool, 4° C	28 days[2]	None
Other Metals		8 oz. glass jar[1]	)combined	Cool, 4° C	180 days[2]	None

[1] ...Teflon Lid Liner

[2] ...To Extraction

[3] ...Pack Tightly

The instruments most often utilized in the field are listed below:

HNU PI-101
Foxboro OVA-108
Foxboro OVA-128
Sampler, Discrete
  Manning
Sampler, Discrete
  Microprocessor,
  Manning
Meter, pH
  Ioananalyzer,
  Portable, Beckman
Meter, pH and Temp.
  Recording Analytical
  Measurements
Meter, pH Recording
  Remote Sampling
  Analytical
  Measurements
Meter, Conductivity
  Geonics
Apparatus, Breathing
  SCBA, MSA
Apparatus, Breathing
  Umbilical, MSA
Apparatus, Breathing
  Umbilical, Survivar
Apparatus, Breathing
  60 min/Biopack
Ultrasonic Level/
  Flow, Manning
Analyzer, Engine
  Exhaust, Chrysler
  Corporation
Fluorometer, High
  Volume, Turner
Flowmeter, Dipper
  Manning
Sampler, Source,
  Stack, Misco
Sampler, Source,
  Stack, Scientific
  Glass
Meter, Watr Current,
  Marsh, McBirney

Fishing Equipment
  Shocker, Coffelt
  Electronics
Locator, Cable
  Magnetic, Brunson
  Instruments
Calibrator, Digital
  0-20 OSL, Kurtz
Calibrator, Digital
  0-15 SLPM, Kurtz
FIT Testing Apparatus,
  Portacount
Calibrator, High
  Volume, Kurtz
Sampler, Discrete
  Microprocessor, ISCO
Flow Meter, ISCO
Flow Metering Inserts,
  ISCO
Geiger Counter Ludlum
Gas Detector, HCN
  Bayer Diagnostic
Gastech, Personal
  Model 6X91
Gastech, Personal
  Model 6X86
Photoionization
  Detector, Micro
  Tip, HL-200
High Volume Air
  Sampler
Low Volume Air
  Sampler
Portable Generator
  EMS4000, Honda
Well Sampler
  Pneumatic Pump
Recorder, Sound
  Ultrasonic, Level
Well Depth Sounder
Conductivity Meter
ORSAT Analyzer
  Stack Gas

Although the above parameter-specific instruments, or their equivalent, should be used almost exclusively, field personnel may occasionally utilize a multi-parameter instrument for the measurement of all the above mentioned parameters with the exception of chlorine residual. The instrument is manufactured by the Hydrolab Corporation and is designed specifically for use in the "field."

Proper calibration of these instruments is considered an essential ingredient of the measurement process to ensure the collection of valid "field" data. In general, these instruments are calibrated according to the manufacturer's recommended procedures. The pH meter, dissolved oxygen meters, and the colorimeters are normally calibrated daily in the field, just prior to use. Calibration of these instruments, according to the manufacturer's specifications, is normally sufficient to ensure the collection of valid data even when taking a number of measurements. However, if in the field inspector's judgment, a drastic change in the field conditions occurs or if an instrument is subjected to other than normal handling, the instrument should be recalibrated. The only exception to the above procedures is that the multi-parameter instrument (Hydrolab) should only be calibrated in a laboratory environment, just prior to being utilized in the "field."

EQUIPMENT DECONTAMINATION

When possible, the investigators should use new disposable sampling utensils such as plastic scoops, stainless steel spatulas, glass colowassas or laboratory cleaned glass jars, since no additional decontamination is needed for this equipment. Nonexpendable equipment must be decontaminated before and after each use. This equipment includes, but is not limited to, shovels, teflon bailers, soil augers (powered and hand operated), soil probes, buckets, automatic samplers, etc. The portion of these sampling devices that come in direct contact with the sample must be washed with a soapy water solution, using a non-phosphate laboratory cleaner and vigorous scrubbing (scrub brush). The equipment may require disassembly to ensure that contamination is removed from all surfaces. Two tap water rinses follow. A third and final rinse should consist of laboratory deionized organic free water. In order to reduce the likelihood of cross-contamination due to equipment, sampling should proceed from cleanest areas first to dirtiest areas last, whenever possible. Between sampling stations, the equipment is decontaminated, as described above. At some point during the sampling effort, deionized organic free lab water should be passed through or over the newly decontaminated sampling equipment and then sampled to ensure that the decontamination procedure was effective. This so-called equipment blank should be preserved, returned to the laboratory using appropriate chain-of-custody procedures, and analyzed for the same parameters as the actual samples. The above decontamination procedure can be modified for specific parameters and conditions, if deemed necessary by the team leader.

Solvents should only be used if proper decontamination cannot be obtained with soapy water (e.g., heavy petroleum products) or if specifically requested. If a solvent is used, the laboratory analyzing the samples should be consulted to ensure the solvent of choice will not interfere with the analytical procedure or mask the results.

SAMPLE LOGGING

A sample log should be maintained in a bound log book which documents all samples that are collected. This log should include a unique sample number (if needed), date, time, sample medium (soil, liquid, etc.), preservative (if any), parameters, and location. Included with the log are any observations made by the sampler that would otherwise identify the sample or conditions at the time of sampling. If photographs are taken, that should also be noted in the field log book.

SAMPLE SPLITTING

Often, it will be necessary to collect duplicate samples or to split a sample in order to provide the facility with a separate sample it can analyze on its own. In these situations, every effort should be made to ensure that both samples are as identical as possible and should theoretically yield the same results. Bulk samples (liquid) for parameters such as extractable organics, cyanide, nutrients, PCB, metals, etc. may be collected in a larger container and alternately poured into the appropriate sample containers. However, the liquid should be well mixed during the transfer. EPA normally provides the sample containers.

Certain parameters may not be split using the above method (e.g., volatile organics, semi-volatiles, and oil and grease) since these samples must be collected and analyzed in their original container. This type of sample should not be distributed by splitting since it may cause air stripping of the volatiles or, in the case of the oil and grease, a residue may adhere to the sample container and cause an erroneous measurement. These parameters will necessitate collecting duplicate samples, virtually at the same time and at the same place to assure homogeneity.

Samples of solid matrices such as sludge, soil, or sediment may be placed into a sufficiently large container and either hand agitated and/or mechanically mixed with a blender, etc. to achieve a homogenous consistency, except for volatile or semi-volatile analyses. Individual samples may then be placed in the appropriate containers.

Wipe samples (PCB), due to their nature, must be collected immediately adjacent to each other with each party receiving a separate sample.

CHAIN-OF-CUSTODY

Since there are legal implications when sampling data is used as evidence, EPA must be able to demonstrate that the samples were protected from tampering from the time of collection to the time they are introduced as evidence. This demonstration is based on samples being in the possession or custody of an EPA employee at all times, and it is documented by means of a chain-of-custody record. Custody implies both physical possession as well as controlled access by locking in a secured area. The chain-of-custody record indicates who had possession at any given point in time and how and when transfer of custody from one individual to another occurred.

Each sample container should be affixed with a sample tag to ensure chain-of-custody from the sample location to the laboratory (see next section for further details). The tag contains the sample location, the sampler's signature, preserved (Y or N), parameter(s), and sample type (composite or grab) and it must be completely filled out using waterproof ink.

Pressure sensitive tape affixed to the container may also be used under certain circumstances to identify the sample. Ink used to write on the container must be waterproof. At a minimum, the label must contain the following information: location, date, time, sample number (if needed), and preservative used. The sampler should be certain that the label is securely affixed to the outside of the container and will not peel off during shipment.

The chain-of-custody record is comprised of sample tags and the record form itself. Both of these are shown on the following pages. A sample tag should be completed for each sample by the field sampler using waterproof ink, if possible. It should be affixed to the sample container in a secure manner. The field sampler should also complete the chain-of-custody record form, appropriately describing all samples that he/she was responsible for collecting. The same wording must be used on both the tag and form, and care must be exercised to make sure that all the information in the chain-of-custody record corresponds properly without discrepancies. While the samples are in his/her custody, all necessary precautions should be taken by the field sampler to ensure that they are adequately safeguarded. Whenever the possession of samples is transferred, the individuals relinquishing and receiving will sign, date, and note the time on the record. The record will continue to accompany the samples. At the completion of the process, a copy of all of the chain-of-custody records will be provided to the team leader for filing.

TRANSPORTATION

All samples will be properly packed in suitable ice chests and transported back to the Regional Laboratory via vehicle or private transport. Chain-of-Custody record forms should also be affixed to the ice chests. The inspectors should always lock the vehicles in which samples are being transported. There may be times when DOT regulations will have to be followed. At least one member of the inspection team should review the DOT requirements prior to an inspection, and make certain that they are complied with in cases where samples are unusually hazardous or travel through tunnels or if confined/special areas are encountered.

UNITED STATES ENVIRONMENTAL PROTECTION AGENCY

NATIONAL ENFORCEMENT INVESTIGATIONS CENTER
Building 53, Box 25227, Denver Federal Center
Denver, Colorado 80225

EPA

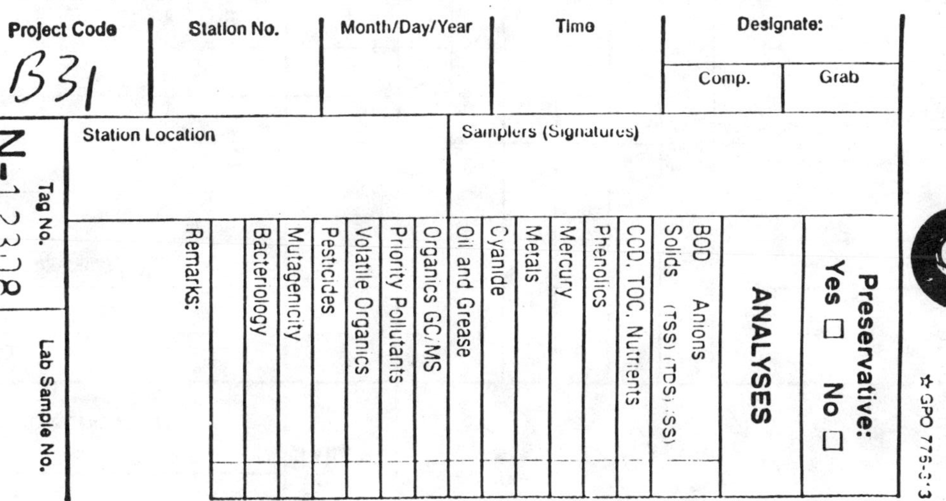

Project Code	Station No.	Month/Day/Year	Time	Designate:	
B31				Comp.	Grab

		Preservative: Yes ☐ No ☐
Tag No. N-12308	Station Location	**ANALYSES**
Lab Sample No.		BOD Anions
		Solids (TSS) (TDS) (SS)
		CCD, TOC, Nutrients
		Phenolics
		Mercury
		Metals
		Cyanide
		Oil and Grease
		Organics GC/MS
		Priority Pollutants
		Volatile Organics
		Pesticides
		Mutagenicity
		Bacteriology
	Remarks:	

Samplers (Signatures)

☆ GPO 776-313

ENVIRONMENTAL PROTECTION AGENCY
Office of Enforcement

NATIONAL ENFORCEMENT INVESTIGATIONS CENTER
Building 53, Box 25227, Denver Federal Center
Denver, Colorado 80225

CHAIN OF CUSTODY RECORD

PROJ. NO.	PROJECT NAME					NO. OF CON-TAINERS						REMARKS
B31												
SAMPLERS: *(Signature)*												
STA. NO.	DATE	TIME	COMP	GRAB	STATION LOCATION							

Relinquished by: *(Signature)*	Date / Time	Received by: *(Signature)*	Relinquished by: *(Signature)*	Date / Time	Received by: *(Signature)*
Relinquished by: *(Signature)*	Date / Time	Received by: *(Signature)*	Relinquished by: *(Signature)*	Date / Time	Received by: *(Signature)*
Relinquished by: *(Signature)*	Date / Time	Received for Laboratory by: *(Signature)*	Date / Time	Remarks	

Distribution: Original Accompanies Shipment, Copy to Coordinator Field Files

APPENDIX N

LAND DISPOSAL RESTRICTIONS PROGRAM

Appendix N
LAND DISPOSAL RESTRICTIONS PROGRAM

BASIC PROGRAM

Land Disposal Restrictions (LDR), 40 CFR 268, are phased regulations prohibiting land disposal of hazardous waste unless that waste meets the applicable treatment standards.[*] Land disposal includes but is not limited to placement in a landfill, surface impoundment, waste pile, injection well, land treatment facility, salt dome formation, salt bed formation, underground mine or cave, or placement in a concrete vault or bunker intended for disposal purposes. The applicable treatment standards are expressed as either concentrations of contaminants in the extract or total waste and as specified technologies.

The schedule for the different groups of waste is:

- Solvents and Dioxins: banned from land disposal (unless treated) effective November 8, 1986 and November 8, 1988, respectively

- "California List" Waste: This group includes liquid wastes containing metals, free cyanides, polychlorinated biphenyls, corrosives (pH less than 2.0) and certain wastes containing halogenated organic compounds. Solid hazardous waste containing halogenated organic compounds are also included. These wastes were banned effective July 8, 1987

- "First, Second, and Third Third" Wastes: The remaining listed and characteristic wastes[**] were divided into thirds (see 40 CFR 268 for specific waste groupings). The first third wastes were banned effective August 8, 1988, second third June 8 1989 and the third third May 8, 1990.

---

[*] *40 CFR 268.40, 41, 42, and 43 contain the treatment standards.*
[**] *40 CFR 261 defines listed and characteristic wastes.*

- Newly Listed Wastes: New wastes that become listed after November 8, 1984 will be banned on a case-by-case basis. There is no statutory deadline for determining treatment standards.

The effective dates for banning these wastes from land disposal can be modified by several kinds of variances. 40 CFR 268, Appendix VII includes all of the different effective dates for each type of waste. Effective dates can be modified by any of the following:

- National capacity variance
- Case-by-case extension
- Treatability variance
- Equivalent method variance
- No-migration petition
- Surface impoundment exemption

Generators of restricted waste are required to:

- Determine whether they generate restricted wastes
- Determine waste treatment standards
- Determine whether waste exceeds treatment standards
- Provide for appropriate treatment and/or disposal
- Satisfy documentation, recordkeeping, notification, certification, packaging, and manifesting requirements
- Meet applicable requirements if the generator is or becomes a TSDF

Treatment, storage, and disposal facilities are required to:

- Ensure compliance with generator recordkeeping requirements when residues generated from treating restricted wastes are manifested off-site

- Certify that treatment standards have been achieved for particular wastes prior to disposal

To become familiar with all of the requirements of LDR, refer to 40 CFR 268 and the Land Disposal Restrictions - Summary of Requirements, OSWER 9934.0-1A, February 1991 for a complete discussion.

EVALUATING COMPLIANCE

LDR requires substantial documentation certifying waste types, required treatment, and notifying waste handlers of the regulatory requirements. Interviews and field observations also may be helpful.

Interviews may cover:

- Who fills out the LDR notifications and certifications
- Frequency of sampling and methods used for sampling and analysis

Documentation required to be kept by a generator include:

- LDR notifications and certifications
- Waste analysis plan if treating a prohibited waste in tanks or containers

Documentation required for TSDFs include:

- Storage

    - Waste analyses and results
    - Waste analysis plan (provision for determining whether a waste is prohibited)

- Treatment

    - Waste analysis plan

- Land Disposal Facility

    - Generator and treatment facility notifications and certifications
    - Waste analysis plan

Field/visual observations related to LDR requirements can be incorporated into a general storage facility inspection. LDR wastes cannot be stored longer than 1 year unless the facility can show that the storage is solely for the purpose of accumulation of sufficient quantities of hazardous waste necessary to facilitate proper recovery, treatment, or disposal. Wastes that are placed in storage prior to the effective date of the restrictions for that waste are not subject to the LDR restrictions on storage. A quick check of accumulation dates on labels will determine how long a drum has been in storage. Be sure to note the hazardous waste number.

Refer to the Land Disposal Restrictions - Inspection Manual, OSWER 9938.1A, February 1989, for a complete discussion on how to conduct an LDR inspection. Keep in mind that this document has yet to be updated with information regarding the Third Third wastes.

APPENDIX O

SOURCES SUBPART (40 CFR PART 60) EFFECTIVE DATE OF
    STANDARD AIR POLLUTANTS SUBJECT TO NSPS

NSPS SOURCES REQUIRING CEM

SOURCES SUBJECT TO TITLE 40 CFR PART 61 NATIONAL EMISSIONS
    STANDARDS FOR HAZARDOUS AIR POLLUTANTS

LIST OF HAZARDOUS AIR POLLUTANTS TO BE REGULATED UNDER
    AIR TOXICS PROGRAM

Table 1

SOURCES SUBPART (40 CFR Part 60)
EFFECTIVE DATE OF STANDARD AND POLLUTANTS SUBJECT TO NSPS

Source	Subpart	Effective Date	Pollutant
Fossil-fuel-fired steam generators constructed after August 17, 1971	D	August 17, 1971	Particulate matter, sulfur dioxide, nitrogen oxides
Fossil-fuel-fired steam generator constructed after September 18, 1978	Da	September 18, 1978	Particulate matter, sulfur dioxide, nitrogen dioxide
Industrial-Commercial-Institutional steam generating units constructed after June 19, 1984	Db	June 19, 1984	Particulate matter, sulfur dioxide, nitrogen oxides
Municipal incinerators	E	August 17, 1971	Particulate matter
Portland cement plants	F	August 17, 1971	Particulate matter
Nitric acid plants	G	August 17, 1971	
Sulfuric acid plants	H	August 17, 1971	Sulfur dioxide, acid mist (sulfuric acid)
Asphalt concrete plants	I	June 11, 1973	Particulate matter
Petroleum refineries	J	June 11, 1973	Particulate matter, carbon monoxide, sulfur dioxide
Storage vessels for petroleum liquids	K	June 11, 1973	VOC
	Ka	May 18, 1978	VOC
Volatile organic liquid storage vessels	Kb	July 23, 1984	VOC
Secondary lead smelters	L	June 11, 1973	Particulate matter
Secondary brass and bronze ingot production plants	M	June 11, 1973	Particulate matter
Iron and steel plants (basic oxygen furnace)	N	June 11, 1973	Particulate matter
Iron and steel plants (secondary emissions from oxygen furnaces)	Na	January 20, 1983	Particulate matter
Sewage treatment plants (incinerators)	O	June 11, 1973	Particulate matter
Primary copper smelters	P	October 16, 1974	Particulate matter, sulfur dioxide
Primary zinc smelters	Q	October 16, 1974	Particulate matter, sulfur dioxide
Primary lead smelters	R	October 16, 1974	Particulate matter, sulfur dioxide
Primary aluminum reduction plants	S	October 23, 1974	Fluorides
Phosphate fertilizer industry (listed as five separate categories)	T U V W X	October 22, 1974	Fluorides
Coal preparation plants	Y	October 24, 1974	Particulate matter
Ferro-alloy production facilities	Z	October 21, 1974	Particulate matter, carbon monoxide
Steel plants (electric arc furnaces)	AA	October 21, 1974	Particulate matter
Steel plants, electric arc furnaces and argon-oxygen decarburization vessels	AAa	August 17, 1983	Particulate matter
Kraft pulp mills	BB	September 24, 1976	Particulate matter, TRS
Glass plants	CC	June 15, 1979	Particulate matter
Grain elevators	DD	August 3, 1978	Particulate matter
Metal furniture surface coating	EE	November 28, 1980	VOC
Stationary gas turbines	GG	September 24, 1976	Nitrogen oxides, sulfur dioxide
Lime plants	HH	May 3, 1977	Particulate matter
Lead acid battery plants	KK	January 14, 1980	Lead
Metallic mineral processing plants	LL	August 24, 1982	Particulate matter
Auto and light-duty truck, surface coating operation	MM	October 5, 1979	VOC
Phosphate rock plants	NN	September 21, 1979	Particulate matter
Ammonium sulfate plants	PP	February 4, 1980	Particulate matter
Graphic arts industry	QQ	October 28, 1980	VOC
Pressure sensitive tape manufacturing	RR	December 30, 1980	VOC
Appliance surface coating	SS	December 24, 1980	VOC
Metal coil surface coating	TT	January 5, 1981	VOC

Table 1 (cont.)

SOURCES SUBPART (40 CFR Part 60)
EFFECTIVE DATE OF STANDARD AND POLLUTANTS SUBJECT TO NSPS

Source	Subpart	Effective Date	Pollutant
Asphalt roofing plants	UU	November 18, 1980; May 26, 1981	Particulate matter
Synthetic organic chemicals	VV	January 5, 1981	Performance standards
Beverage can surface coating	WW	November 26, 1980	VOC
Bulk gasoline terminal	XX	December 17, 1980	VOC
New residual wood heaters	AAA	July 1, 1988	Particulate matter
Rubber tire manufacturing industry	BBB	January 20, 1983	VOC
Vinyl/urethane coating	FFF	January 18, 1983	VOC
Petroleum refineries	GGG	January 4, 1983	Performance standards
Synthetic fiber plants	HHH	November 23, 1982	VOC
Synthetic organic chemicals (air oxidation unit processes)	III	October 21, 1983	VOC
Petroleum dry cleaners	JJJ	September 21, 1984	VOC
Onshore natural gas processing plants	KKK	June 24, 1985	VOC
Onshore natural gas processing plants	LLL	October 1985	SO_2
Synthetic organic chemicals (distillation operations)	NNN	December 30, 1983	VOC
Nonmetallic mineral processing plants	OOO	August 1, 1985	Particulate matter
Wool fiberglass insulation manufacturing plants	PPP	February 25, 1985	Particulate matter
Petroleum refineries (wastewater systems)	QQQ	May 4, 1987	VOC
Magnetic tape coating	SSS	January 22, 1986	VOC
Industrial surface coating, plastic parts for business machines	TTT	January 8, 1986	VOC
Polymeric coating of supporting substrates facilities	VVV	April 30, 1987	VOC

Table 2

NSPS SOURCES REQUIRING CEM

Source	Subpart	Effective Date	Monitor
Fossil-fuel-fired steam generator	D	08/17/71	opacity, SO_2, NO_x, O_2 or CO_2
Fossil-fuel-fired electric utilities	Da	09/18/78	opacity, SO_2, NO_x, O_2 or CO_2
Nitric acid plants	G	08/17/71	NO_x
Sulfuric acid plants	H	08/17/71	SO_2
Petroleum refineries (FBCCU)	J	06/11/73	opacity, CO, SO_2, H_2S
Claus sulfur recovery unit	J	10/04/76	opacity, CO, SO_2, H_2S
Primary copper smelters	P	10/16/74	opacity, SO_2
Primary zinc smelters	Q	10/16/74	opacity, SO_2
Primary lead smelters	R	10/16/74	opacity, SO_2
Ferroalloy production facilities	Z	10/21/74	opacity
Electric arc furnaces	A A	10/21/74	opacity
Kraft pulp mills	BB	09/24/76	opacity, TRS
Lime manufacturing plants	HH	05/03/77	opacity
Phosphate rock plants	NN	09/21/79	opacity
Flexible vinyl and urethane coating and printing	FFF	01/18/83	VOC
Onshore natural gas processing plants	LLL	10/01/85	SO_2/T/TRS

Table 3
SOURCES SUBJECT TO TITLE 40 CFR PART 61
NATIONAL EMISSIONS STANDARDS FOR HAZARDOUS AIR POLLUTANTS

Pollutant	Subpart	Source
Radon-222	B	Underground uranium mines
Beryllium	C	Extraction plants Ceramic plants Foundries Incinerators Machine shops
Beryllium	D	Rocket motor firing
Mercury	E	Ore processing plants Chlor-alkali plants Sludge incinerators Sludge drying plants
Vinyl chloride	F	Ethylene dichloride plants Vinyl chloride plants Polyvinyl chloride plants
Radionuclides	H	DOE facilities
Radionuclides	I	Facilities licensed by the Nuclear Regulatory Commission and Federal facilities not covered by Subpart H
Benzene (leaks)	J	Equipment in benzene service (plants designed to produce more than 1,000 megagrams of benzene per year)
Radionuclides	K	Elemental phosphorus plants
Benzene	L	Coke by-product recovery plants
Asbestos	M	Asbestos mills Manufacturing Demolition and renovation Spraying Fabrication Waste disposal
Inorganic arsenic	N	Glass manufacturing plants
Inorganic arsenic	O	Primary copper smelters
Inorganic arsenic	P	Arsenic trioxide and metallic arsenic production facilities

Table 3 (cont.)

SOURCES SUBJECT TO TITLE 40 CFR PART 61
NATIONAL EMISSIONS STANDARDS FOR HAZARDOUS AIR POLLUTANTS

Pollutant	Subpart	Source
Radon-222	Q	DOE facilities
Radon-222	R	Phosphogypsum stacks
Radon-222	T	Disposal sites of uranium mill tailings
Volatile hazardous air (VHAP)*	V	Equipment leaks (fugitive pollutants emission sources)
Radon-222	W	Licensed uranium mill tailings
Benzene	Y	Benzene storage vessels
Benzene	BB	Benzene transfer operations
Benzene	FF	Benzene waste operations

* *Volatile hazardous air pollutant (VHAP) means a substance regulated under this part for which a standard for equipment leaks has been proposed and promulgated. As of February 1, 1989, benzene and vinyl chloride are VHAPs.*

Table 4

LIST OF HAZARDOUS AIR POLLUTANTS TO BE REGULATED UNDER AIR TOXICS PROGRAM

CAS Number	Chemical name	CAS Number	Chemical Name
75070	Acetaldehyde	67663	Chloroform
60355	Acetamide	107301	Chloromethyl methyl ether
75058	Acetonitrile	126998	Chloroprene
98862	Acetophenone	1319773	Cresols/cresylic acid (isomers and mixture)
53963	2-Acetylaminofluorene	95487	o-Cresol
107028	Acrolein	108394	m-Cresol
79061	Acrylamide	106445	p-Cresol
79107	Acrylic acid	98828	Cumene
107131	Acrylonitrile	93747	2,4-D, salts and esters
107051	Allyl chloride	3547044	DDE
92671	4-Aminobiphenyl	334883	Diazomethane
62533	Analine	132649	Dibenzofurans
90040	o-Anisidine	96128	1,2-Dibromo-3-chloropropane
1332214	Asbestos	84742	Dibutylphthalate
71432	Benzene (including benzene from gasoline)	106467	1,4-Dibhlorobenzene(p)
92875	Benzidine	91941	3,3-Dichlorobenzidene
98077	Benzotrichloride	111444	Dichloroethyl ether [Bis(2-chloroethyl) ether]
100447	Benzyl chloride	542756	1,3-Dichloropropene
92524	Biphenyl	62737	Dichlorvos
117817	Bis(2-ethylhexyl)phthalate (DEHP)	111422	Diethanolamine
542881	Bis(chloromethyl)ether	121697	N,N-Diethyl aniline (N,N-Dimethylaniline)
75252	Bromoform	64675	Diethyl sulfate
106990	1,3-Butadiene	119904	3,3-Dimethoxybenzidine
156627	Calcium cyanamide	60117	Dimethyl aminoazobenzene
105602	Caprolactam	119937	3,3'-Dimethyl benzidine
133062	Captan	79447	Dimethyl carbamoyl chloride
63252	Carbaryl	68122	Dimethyl formamide
75150	Carbon disulfide	57147	1,1-Dimethyl hydrazine
56235	Carbon tetrachloride	13113	Dimethyl phthalate
463581	Carbonyl sulfide	77781	Dimethyl sulfate
120809	Catechol	534521	4,6-Dinitro-o-cresol
133904	Chloramben	51285	2,4-Dinitrophenol
57749	Chlordane	121142	2,4-Dinitrotoluene
7782505	Chlorine	123911	1,4-Dioxane(1,4-Diethyleneoxide)
79118	Chloroacetic acid	122667	1,2-Diphenyl hydrazine
532274	2-Chloroacetophenone	106898	Epichlorohydrin(1-Chloro-2,3-epoxypropane)
108907	Chlorobenzene	106887	1,2-Epoxybutane
510156	Chlorobenzilate	140885	Ethyl acrylate

Table 4(cont.)

LIST OF HAZARDOUS AIR POLLUTANTS TO BE REGULATED UNDER AIR TOXICS PROGRAM

CAS Number	Chemical name	CAS Number	Chemical Name
100414	Ethyl benzene	80626	Methyl methacrylate
51796	Ethyl carbamate (Urethane)	101144	4,4-Methylene bis(2-chloroaniline)
75003	Ethy chloride (Chloroethane)	75092	Methylene chloride (Dichloromethane)
106934	Ethylene dibromide (Dibromoethane)	101688	Methylene diphenyl diisocyanate (MDI)
107062	Ethylene dichloride (1,2-Dichloroethane)	101779	4,4'-Methylenedianiline
107211	Ethylene glycol	1634044	Methyl tert-butyl ether
15164	Ethylene iminie (Aziridine)	91203	Naphthalene
75218	Ethylene oxide	98953	Nitrobenzene
96457	Ethylene thiourea	92933	4-Nitrobiphenyl
75343	Ethylidene dichloride (1,1-Dichloroethane)	100027	4-Nitrophenol
50000	Formaldehyde	79469	2-Nitropropane
76448	Heptachlor	684935	N-Nitroso-N-methylurea
118741	Hexachlorobenzene	62759	N-Nitrosodimethylamine
87683	Hexachlorobutadiene	59892	N-Nitrosomorpholine
77474	Hexachlorocyclopentadiene	56382	Parathion
67721	Hexachloroethane	82688	Pentachloronitrobenzene (Quintobenzene)
822060	Hexamethylene-1,6-diisocyanate	87865	Pentachlorophenol
680319	Hexamethylphosphoramide	108952	Phenol
110543	Hexane	106503	p-Phenylenediamine
302012	Hydrazine	75445	Phosgene
7647010	Hydrochloric acid	7803512	Phosphine
7664393	Hydrogen fluoride (Hydrofluoric acid)	7723140	Phosphorus
7783064	Hydrogen sulfide[6]	85449	Phthalic anhydride
123319	Hydroquinone	1336363	Polychlorinated biphenyls (Aroclors)
78591	Isophorone	1120714	1,3-Propane sultone
58899	Lindane (all isomers)	57578	beta-Propiolactone
108316	Maleic anhydride	123386	Propionaldehyde
67561	Methanol	114261	Propoxur (Baygon)
72435	Methoxychlor	78875	Propylene dichloride (1,2-Dichloropropane)
74839	Methyl bromide (Bromomethane)	75569	Propylene oxide
74873	Methyl chloride (Chloromethane)	75558	1,2-Propylenimine (2-Methyl aziridine)
71556	Methyl chloroform (1,1,1-Trichloroethane)		Quinoline
78933	Methyl ethyl ketone (2-Butanone)	106514	Quinone
60344	Methyl hydrazine	100425	Styrene
74884	Methyl iodide (Iodomethane)	96093	Styrene oxide
108101	Methyl isobutyl ketone (Hexone)	1746016	2,3,7,8-Tetrachlorodibenzo-p-dioxin
624839	Methyl isocyanate	79345	1,1,2,2-Tetrachloroethane

Table 4 (cont.)

LIST OF HAZARDOUS AIR POLLUTANTS TO BE REGULATED UNDER AIR TOXICS PROGRAM

CAS Number	Chemical name	CAS Number	Chemical Name
127184	Tetrachloroethylene (Perchloroethylene)		Antimony compounds*
7550450	Titanium tetrachloride		Arsenic Compounds (inorganic including arsine)*
108883	Toluene		Beryllium compounds*
95807	2,4-Toluene diamine		Cadmium compounds*
584849	2,4-Toluene diisocyanate		Chromium compounds*
95534	o-Toluidine		Cobalt compounds*
8001352	Toxaphene (Chlorinated camphene)		Coke oven emissions
120821	1,2,4-Trichlorobenzene		Cyanide Compounds*[1]
79005	1,1,2-Trichloroethane		Glycol ethers*[2]
79016	Trichloroethylene		Lead Compounds*
95954	2,4,5-Trichlorophenol		Manganese Compounds*
88062	2,4,6-Trichlorophenol		Mercury Compounds*
121448	Triethylamine		Fine mineral fibers[3]
1582098	Trifluralin		Nickel compounds*
540841	2,2,4-Trimethylpentane		Polycrylic Organic Matter[4]
108054	Vinyl acetate		Radionuclides (including radon)[5]
593602	Vinyl bromide		Selenium Compounds*
75014	Vinyl chloride		
75354	Vinylidene chloride (1,1-Dichloroethylene)		
1330207	Xylenes (isomers and mixtures)		
95476	o-Xylenes		
108383	m-Xylenes		
106423	p-Xylenes		

*   *Unless otherwise specified, these listings are defined as including any unique chemical substance that contains the named chemical (i.e., antimony, arsenic, etc.) as part of the chemical's infrastructure.*

APPENDIX P

TSCA FORMS

Appendix P
CONFIDENTIALITY NOTES AND DISCUSSIONS

The TSCA Notice of Inspection [Figure L-1] and Inspection Confidentiality Notice [Figure L-2] are presented to the facility owner or agent in charge during the opening conference. These notices inform facility officials of their right to claim as confidential business information any information (documents, physical samples, or other material) collected by the inspector.

<u>Authority to Make Confidentiality Claims</u>

The inspector must ascertain whether the facility official, to whom the notices were given, has the authority to make business confidentiality claims for the company. The facility official's signature must be obtained at the appropriate places on the notices certifying that he does or does not have such authority.

- The facility owner is assumed to always have the authority to make business confidentiality claims. In most cases, it is expected that the agent in charge will also have such authority. It is possible that the officials will want to consult with their attorneys (or superiors in the case of agents in charge) regarding this issue.

- If no one at the site has the authority to make business confidentiality claims, a copy of the TSCA Inspection Confidentiality Notice and Notice and Declaration of Confidential Business Information form [Figure L-3] are to be sent to the chief executive officer of the firm within 2 days of the inspection. He will then have 7 calendar days in which to make confidentiality claims.

- The facility official may designate a company official, in addition to the chief executive officer, who should also receive a copy of the notices and any accompanying forms.

☣EPA

US ENVIRONMENTAL PROTECTION AGENCY
WASHINGTON, DC 20460

TOXIC SUBSTANCES CONTROL ACT

NOTICE OF INSPECTION

Form Approved
OMB No 2070-0007
Expires 3-31-88

1. INVESTIGATION IDENTIFICATION			2. TIME	3. FIRM NAME
DATE	INSPECTOR NO.	DAILY SEQ. NO.		

4. INSPECTOR ADDRESS	5. FIRM ADDRESS

REASON FOR INSPECTION

Under the authority of Section 11 of the Toxic Substances Control Act :

☐ For the purpose of inspecting (including taking samples, photographs, statements, and other inspection activities) an establishment, facility, or other premises in which chemical substances or mixtures or articles containing same are manufactured, processed or stored, or held before or after their distribution in commerce (including records, files, papers, processes, controls, and facilities) and any conveyance being used to transport chemical substances, mixtures, or articles containing same in connection with their distribution in commerce (including records, files, papers, processes, controls, and facilities) bearing on whether the requirements of the Act applicable to the chemical substances, mixtures, or articles within or associated with such premises or conveyance have been complied with.

☐ In addition, this inspection extends to *(Check appropriate blocks)*:

    ☐ A. Financial data   ☐ D. Personnel data

    ☐ B. Sales data   ☐ E. Research data

    ☐ C. Pricing data

The nature and extent of inspection of such data specified in A through E above is as follows:

A42-20

INSPECTOR SIGNATURE	RECIPIENT SIGNATURE	
NAME	NAME	
TITLE	DATE SIGNED	TITLE	DATE SIGNED

EPA Form 7740-3 (12-82)

US ENVIRONMENTAL PROTECTION AGENCY
WASHINGTON, DC 20460

♻EPA

TOXIC SUBSTANCES CONTROL ACT

TSCA INSPECTION CONFIDENTIALITY NOTICE

Form Approved
OMB No 2070-0007
Expires 3-31-88

1. INVESTIGATION IDENTIFICATION			2. FIRM NAME
DATE	INSPECTOR NO.	DAILY SEQ. NO.	

3. INSPECTOR NAME

4. FIRM ADDRESS

5. INSPECTOR ADDRESS

6. CHIEF EXECUTIVE OFFICER NAME

7. TITLE

TO ASSERT A CONFIDENTIAL BUSINESS INFORMATION CLAIM

It is possible that EPA will receive public requests for release of the information obtained during inspection of the facility above. Such requests will be handled by EPA in accordance with provisions of the Freedom of Information Act (FOIA), 5 USC 552; EPA regulations issued thereunder, 40 CFR Part 2; and the Toxic Substances Control Act (TSCA), Section 14. EPA is required to make inspection data available in response to FOIA requests unless the Administrator of the Agency determines that the data contain information entitled to confidential treatment or may be withheld from release under other exceptions of FOIA.

Any or all the information collected by EPA during the inspection may be claimed confidential if it relates to trade secrets or commercial or financial matters that you consider to be confidential business information. If you assert a CBI claim, EPA will disclose the information only to the extent, and by means of the procedures set forth in the regulations (cited above) governing EPA's treatment of confidential business information. Among other things, the regulations require that EPA notify you in advance of publicly disclosing any information you have claimed as confidential business information.

A confidential business information (CBI) claim may be asserted at any time. You may assert a CBI claim prior to, during, or after the information is collected. The declaration form was developed by the Agency to assist you in asserting a CBI claim. If it is more convenient for you to assert a CBI claim on your own stationery or by marking the individual documents or samples "TSCA confidential business information," it is not necessary for you to use this form. The inspector will be glad to answer any questions you may have regarding the Agency's CBI procedures.

While you may claim any collected information or sample as confidential business information, such claims are unlikely to be upheld if they are challenged unless the information meets the following criteria:

1.   Your company has taken measures to protect the confidentiality of the information, and it intends to continue to take such measures.

2.   The information is not, and has not been, reasonably obtainable without your company's consent by other persons (other than governmental bodies) by use of legitimate means (other than discovery based on showing of special need in a judicial or quasi-judicial proceeding).

3.   The information is not publicly available elsewhere.

4.   Disclosure of the information would cause substantial harm to your company's competitive position.

At the completion of the inspection, you will be given a receipt for all documents, samples, and other materials collected. At that time, you may make claims that some or all of the information is confidential business information.

If you are not authorized by your company to assert a CBI claim, this notice will be sent by certified mail, along with the receipt for documents, samples, and other materials to the Chief Executive Officer of your firm within 2 days of this date. The Chief Executive Officer must return a statement specifying any information which should receive confidential treatment.

The statement from the Chief Executive Officer should be addressed to:

and mailed by registered, return-receipt requested mail within 7 calendar days of receipt of this Notice. Claims may be made any time after the inspection, but inspection data will not be entered into the special security system for TSCA confidential business information until an official confidentiality claim is made. The data will be handled under the agency's routine security system unless and until a claim is made.

A42-20

TO BE COMPLETED BY FACILITY OFFICIAL RECEIVING THIS NOTICE:	If there is no one on the premises of the facility who is authorized to make business confidentiality claims for the firm, a copy of this Notice and other inspection materials will be sent to the company's chief executive officer. If there is another company official who should also receive this information, please designate below.
I have received and read the notice	
SIGNATURE	NAME
NAME	TITLE
TITLE DATE SIGNED	ADDRESS

EPA Form 7740-4 (12-82)

EPA

US ENVIRONMENTAL PROTECTION AGENCY
WASHINGTON, DC 20460

TOXIC SUBSTANCES CONTROL ACT

DECLARATION OF CONFIDENTIAL BUSINESS INFORMATION

Form Approved
OMB No 2070-000
Expires 3-31-88

1. INVESTIGATION IDENTIFICATION			2. FIRM NAME
DATE	INSPECTOR NO.	DAILY SEQ. NO.	

3. INSPECTOR ADDRESS	4. FIRM ADDRESS

INFORMATION DESIGNATED AS CONFIDENTIAL BUSINESS INFORMATION

NO.	DESCRIPTION

ACKNOWLEDGEMENT BY CLAIMANT

The undersigned acknowledges that the information described above is designated as Confidential Business Information under Section 14(c) of the Toxic Substances Control Act. The undersigned further acknowledges that he/she is authorized to make such claims for his/her firm.

The undersigned understands that challenges to confidentiality claims may be made, and that claims are not likely to be upheld unless the information meets the following guidelines: (1) The company has taken measures to protect the confidentiality of the information and it intends to continue to take such measures; (2) The information is not, and has not been reasonably attainable without the company's consent by other persons (other than governmental bodies) by use of legitimate means (other than discovery based on a showing of special need in a judicial or quasi-judicial proceeding); (3) The information is not publicly available elsewhere; and (4) Disclosure of the information would cause substantial harm to the company's competitive position.

A42-20.

INSPECTOR SIGNATURE	CLAIMANT SIGNATURE
NAME	NAME
TITLE	DATE SIGNED	TITLE	DATE SIGNED

INSPECTION FILE

Appendix P (cont.)

Confidentiality Discussion

Officials should be informed of the procedures and requirements that EPA must follow in handling TSCA confidential business information. The inspector should explain that these procedures were established to protect the companies subject to TSCA and cover the following points during the discussion.

- Data may be claimed confidential business information during the closing conference if a person authorized to make such claims is on-site at the facility.

- It is suggested that a company official accompany the inspector during the inspection to facilitate designation (or avoidance, if possible) of confidential business data.

- A detailed receipt for all documents, photographs, physical samples, and other materials [Figure L-4] collected during the inspection will be issued at the closing conference.

- An authorized person may make immediate declarations that some or all the information is confidential business information. This is done by completing the Declaration of Confidential Business Information form. Each item claimed must meet all four of the criteria shown on the TSCA Inspection Confidentiality Notice.

- If no authorized person is available on-site, a copy of the notices, along with the Receipt for Samples and Documents, will be sent by certified, return-receipt-requested mail to the Chief Executive Officer of the firm and to another company official, if one has been designated.

⊕EPA

US ENVIRONMENTAL PROTECTION AGENCY
WASHINGTON, DC 20460

TOXIC SUBSTANCES CONTROL ACT

RECEIPT FOR SAMPLES AND DOCUMENTS

Form Approved.
OMB No. 2070-0007
Approval expires 3-31-88

1. INVESTIGATION IDENTIFICATION			2. FIRM NAME
DATE	INSPECTOR NO.	DAILY SEQ. NO.	

3. INSPECTOR ADDRESS	4. FIRM ADDRESS

The documents and samples of chemical substances and/or mixtures described below were collected in connection with the administration and enforcement of the Toxic Substances Control Act.

RECEIPT OF THE DOCUMENT(S) AND/OR SAMPLE(S) DESCRIBED IS HEREBY ACKNOWLEDGED:

NO.	DESCRIPTION

A42-

OPTIONAL:

DUPLICATE OR SPLIT SAMPLES: REQUESTED AND PROVIDED ☐   NOT REQUESTED ☐

INSPECTOR SIGNATURE	RECIPIENT SIGNATURE	
NAME	NAME	
TITLE	DATE SIGNED	TITLE	DATE SIGNED

EPA Form 7740-1 (12-82)

Appendix P (cont.)

Four copies are made of the Declaration of Confidential Information form and distributed to:

- Facility owner or agent in charge
- Other company official (if designated)
- Document Control Officer
- Inspection report

APPENDIX Q

TSCA SECTIONS 5 AND 8 GLOSSARY OF TERMS AND ACRONYMS

Appendix Q
GLOSSARY OF TERMS AND ACRONYMS:
TOXIC SUBSTANCES CONTROL ACT SECTIONS 5 AND 8

SECTION 5. New Chemicals

(Note: TSCA does not regulate chemicals such as pesticides, drugs, cosmetics, explosives, etc., which are regulated under separate acts.)

PMN	<u>Premanufacture Notification</u> required for all "new" TSCA chemicals (i.e., those not listed as TSCA Chemical Substances Inventory).
SNURs	<u>Significant New Use Rules</u> require subsequent notification to EPA when usage/exposure of existing chemical changes.
NOC	<u>Notice of Commencement</u> to Agency is required before manufacture begins (after PMN review period has expired).
TME	<u>Test Marketing Exemption</u> from PMN requirement can be obtained on application to and approval by EPA - usually subject to specific restrictions.
LVE	<u>Low Volume Exemption</u> from PMN requirement
PE	<u>Polymer Exemption</u> from PMN requirement, a modified PMN
R & D	<u>Research and Development Exemption</u> - automatic exemption, does not require Agency review or approval.
Section 5(e)	An administrative order limiting the manufacture, Order processing, distribution, use and/or disposal of a chemical for which a PMN is required because there is <u>insufficient informa-tion</u> to permit full evaluation.
Bona fide	Inquiry to Agency to determine whether a chemical is on the confidential portion of the Inventory. A *Bona Fide* should indicate an interest or intent to commercially manufacture the subject chemical.

Appendix Q (cont.)

Section 5(f) An administrative order or rule prohibiting/limiting the
Order/Rule   manufacture, etc., of a chemical for which a PMN is
             required because there is a <u>reasonable basis</u> to conclude that
             such activities present an unreasonable risk to
             health/environment.

Section 8   Existing Chemicals

PAIR
<u>Preliminary Assessment Information Rules</u> are promulgated under Section 8(a) Level A and require reporting to Agency of production, uses and exposure of specific chemicals or classes of chemicals.

ITC
<u>Interagency Testing Committee</u> designates chemicals to be listed in PAIR rules, as well as some of the chemicals in Section 8(d) rules. ITC is established under Section 4(e) of TSCA. It also recommends chemicals for inclusion in testing rules under Section 4(a).

CHEMICAL SUBSTANCES INVENTORY
A listing compiled under Section 8(b) of TSCA of all chemicals manufactured/ processed in U.S. that were manufactured, imported or processed in the period 1975-77. Chemicals for which PMN is submitted are added to inventory when manufacturing/processing commences (i.e., upon receipt of NOC). A major updating of the inventory was undertaken in 1986, and will be repeated every 4 years thereafter.

CAIR
<u>Comprehensive Assessment Information Rule</u>, a more detailed reporting rule under Section 8(A) Level A (see PAIR)

About Government Institutes

Government Institutes, Inc. was founded in 1973 to provide continuing education and practical information for your professional development. Specializing in environmental, health and safety concerns, we recognize that you face unique challenges presented by the ever-increasing number of new laws and regulations and the rapid evolution of new technologies, methods and markets.

Our information and continuing education efforts include a Videotape Distribution Service, over 140 courses held nation-wide throughout the year, and over 150 publications, making us the world's largest publisher in these areas.

Government Institutes, Inc.
4 Research Place, Suite 200
Rockville, MD 20850
(301) 921-2300

Other related books published by Government Institutes:

RCRA Inspection Manual, 2nd Edition - Includes the actual inspection procedures, priorities and checklists EPA will have in hand when they inspect your facility. Covers all details of a RCRA inspection, from facility entry to the closing discussion, including a synopsis of the regulations and the corresponding inspection procedures. Softcover/370 pages/1989/$65 ISBN: 0-86587-762-9

Pesticide Inspection Manual - Written for EPA inspectors conducting pesticides inspections, this manual provides all of the necessary guidance to carry out the standard field procedures. Softcover/231 pages/1989/$65 ISBN: 0-86587-784-X

Clean Water Act Compliance/Enforcement Guidance Manual - Includes a wealth of sample letters, notices, checklists, and forms can be found in chapters on inspections, recordkeeping and reporting, documenting evidence, criminal enforcement, and protecting confidential business information. Softcover/572 pages/1986/$78 ISBN: 0-86587-596-3

NPDES Compliance Inspection Manual, 2nd Edition - Learn to spot the same potential problems that EPA inspectors look for, and correct them before the inspector arrives. A must for those who need to comply with the NPDES program. Get EPA's requirements for recordkeeping, reporting, facility site review, sampling, and more. Softcover/234 pages/1988/$59 ISBN: 0-86587-751-3

TSCA Inspection Manual, Part I - Covers EPA inspectors' authorities and responsibilities; TSCA inspection procedures; post-inspection activities; special procedures; forms; samples; and PCB enforcement program. Softcover/341 pages/1982/$62 ISBN: 0-86587-056-X

TSCA Inspection Manual, Part II - Explains the complex details of EPA's PMN inspection, CFC inspection, Dioxin inspection, TSCA Level A inspection, Asbestos inspection, and Asbestos in Schools inspection. Softcover/216 pages/1986/$58 ISBN: 0-86587-143-4

Call the above number for our current book/video catalog and course schedule.